*Climate Anxiety and
the Kid Question*

Climate Anxiety and the Kid Question

DECIDING WHETHER TO HAVE CHILDREN
IN AN UNCERTAIN FUTURE

Jade S. Sasser

UNIVERSITY OF CALIFORNIA PRESS

University of California Press
Oakland, California

© 2024 by Jade S. Sasser

Cataloging-in-Publication data is on file at the Library of Congress.

ISBN 978-0-520-39382-0 (pbk. : alk. paper)
ISBN 978-0-520-39384-4 (ebook)

Manufactured in the United States of America

33 32 31 30 29 28 27 26 25 24
10 9 8 7 6 5 4 3 2 1

This book is dedicated to Soraiya and Airyssa,
Olive and June
And their parents
And also to my students, past and present, who
have taught me so much

Contents

Acknowledgments

I have wanted to write this book for a long time; my deepest thanks to those who made it possible. To my editors, Michelle Lipinski and Naja Pulliam Collins, and my former editor, Stacy Eisenstark, your insights made it so much sharper and clearer. Thanks also to the three anonymous peer reviewers for their close readings and painstaking attention to detail.

My writing groups have sustained me for so long; thanks to Rajani Bhatia, Ellen Foley, and Emily Merchant for coming along for the ride, as well as to Kalina Michalska, Annika Speer, and Alejandra Dubcovsky for the emotional support. Special thanks to Victoria Reyes, who was there every step of the way. And thanks are also due to the members of the Race/Affect/Environment Colloquium, specifically Britt Wray, Dan Suarez, Michael Méndez, Jaskiran Dhillon, Milanika Turner, and Ingrid Nelson for the generative conversations and the insights on an early chapter draft. Britt, so much of my perspective on this was shaped by our ongoing conversations—much gratitude to you. Many thanks to Amara Miller and her research team. Blanche Verlie and Sarah Jaquette Ray: so much gratitude for our ongoing conversations and collaborative work. I've learned so much from the two of you, even in the midst of a cascading tech fail disaster!

The campus resources and support of the University of California–Riverside made all the difference, including a generous Mellon Dean's Fellowship as well as time and writing space at the Center for Ideas and Society. Although I started there intending to write a completely different book, Covid-19 altered everything and brought this one fortuitously to the forefront. Friends and colleagues whose pep talks, coffees, and writing retreat support made all the difference: Jane Ward, Brandon Robinson, Crystal Baik, Myisha Cherry, Ademide Adelusi-Adeluyi, Jody Benjamin, Liz Przybylski, Donatella Galella, Denise Davis, thank you. Chikako Takeshita, Juliann Allison, Dana Simmons, Ellen Reese, and all of my new colleagues in Society, Environment, and Health Equity, thank you for creating a space to carry these ideas and more into the future.

This book would not exist without the time and generosity of those who participated in interviews, opening their hearts and minds to me. Thanks in particular to Meghan, Josephine, Blythe, Jess, and Emma for the ongoing dialogues. I hope I got it right.

Finally, to the students of GSST 191C in Winter 2023: the day you all collectively melted down in a giant pool of climate-and-reproductive anxiety was a turning point. Thank you for trusting me that day.

Preface

In September 2020, I was in an online meeting when a colleague mentioned that friends were texting her to ask whether her family was planning to evacuate. "Evacuate from what?" I asked, confused. This colleague lived just a few miles from me; surely, if there was something to evacuate from, I would have heard about it too.

Yes, I'd gone out that morning to find a light orange–colored tint in the air, along with the vague smell of smoke. And yes, after a short trip to the grocery store, I found a thin coating of ash covering my car. These were strange, curious events, but I hadn't connected the two. After all, it was 2020: we had experienced months of a terrifying, deadly pandemic; the largest racial justice uprisings in decades; and endless reports of deaths, illnesses, unemployment, and a general sense of upheaval and emotional exhaustion. There was no way I might *also* have to evacuate my home because of a wildfire, *right?!* But here was my colleague on-screen, detailing the plans she and her partner were making to secure their home and child, and depart if necessary. I did a quick internet search and discovered that the

Bobcat fire had broken out several miles away and was beginning to spread. Much as I would have loved to remain in my comfortable state of disbelief, this was not a drill. It was really happening.

I panicked. My heart raced, my breath became shallow, and I felt cold all over. Wild-eyed, I looked around, trying to take a mental calculation of what I would need to pack, where I would go, and how quickly I could do it. Outside, the sky was strangely clear and blue, but it didn't stay that way for long. The greater Los Angeles area soon became enveloped in a thick, murky haze. In my neighborhood, Pasadena, you couldn't see street signs from twenty feet away. It smelled like a campfire at all times—even indoors. There was no way to escape it.

As I packed up to evacuate, I thought, *What would I do if I had children to take care of? Would my worry and fear overwhelm me? How would I parent through this crisis?* I couldn't help but think about my research—at that point I'd spent a couple of years reading about women in their twenties and thirties who were publicly declaring their resistance to having children because of climate change. Suddenly their words made sense, with startling clarity. They connected the impacts of climate change—disrupted air quality, scarce water, extreme temperatures, soil that can't adequately grow nutritious food, loss of species diversity, and the likelihood of extreme weather and wildfires—to the prospect of a livable future. They were uncertain about whether bringing kids into that future was a good idea.

Ranging from eighteen years old to their mid-thirties, these women had several things in common: they were all feeling deeply anxious, sad, and fearful about the state of the planet and the future; they placed the blame for climate change squarely on political leaders who are failing to act aggressively to move away from fossil fuel–based economies; and they saw climate change as an existential threat demanding that they rethink whether having kids is possible. Although climate change is far from the first or only existential threat human communities have faced, it poses the most comprehensive and farthest reaching challenges to the systems that sustain life on Earth. It is a matter of scale, and the scale of this problem is not one that we as a human species have collectively faced before.

When I first encountered the articles about these women who were worried about having kids in the midst of climate change, I didn't connect them to my personal experience in any way.[1] I just thought they could spark interesting research questions. For example, why focus on reproductive questions, when there are so many other, more obvious ways to think about climate impacts? And if you're concerned about having children, why focus on climate change, when there are so many other reasons to worry about being a parent? Were these just privileged young women who'd never had any reason to be worried about the future—much less the present?

But then the Bobcat fire happened, and I began to see the climate-children connection in a new way: sure, it's about how you're feeling in the present, how worried and afraid you might be in response to fires/storms/heat waves, and a general sense of uncertainty about what will happen next. But it's also very much about the future. Pregnancy, childbirth, adoption—having and raising children—while these are lived concerns in the present, they symbolize futures where we can feel good about parenting children, giving them a good life, and leaving some sort of legacy. For many people of reproductive age, that hope is being threatened by climate change.

As a Black woman, I see hope for the future as a tricky thing. I've always been deeply ambivalent about the possibility of parenting a Black child in a country where racism shapes access to such basic rights as health care, education, and the ability to come away from a police encounter alive. My own reproductive years have been filled with unanswered questions about whether, when, how, and why I would have children in a society and world that are profoundly broken. Climate change hasn't been one of my main reasons for concern, but it makes sense that it is driving the reproductive questions of people younger than me. On the one hand, we are collectively dealing with deep and far-ranging changes to our atmosphere, ecological systems, and all of the natural processes that sustain life on

Earth—and our political leaders are *knowingly making it worse*. On the other hand, the impacts of these changes land hardest on the most vulnerable members of society: low-income communities, communities of color, and particularly the youngest members of society, who will experience the effects of climate change throughout their lifetimes. Climate impacts are *profoundly* unjust.

Ironically, the polls, newspaper pieces, and academic journal articles I've read, which focus on this climate-kid connection, have completely ignored the issues of race and class injustice that were so obviously apparent to me. How is this possible? Is race not a part of the picture? How could it not be, given the ways systemic racism and class inequality in the United States leave communities of color significantly more vulnerable to the impacts of climate change in our neighborhoods, bodies, and even our psyches? If more and more people have been writing about how climate change is shaping reproductive questions, why are they leaving people of color out of the discussion?

This book addresses these questions and more: it is about how climate change is shaping the reproductive plans of a growing number of people in the United States and the role of race in that process. It prioritizes the perspectives of Millennials and members of Generation Z, who are actively wrestling with these concerns while trying to plan their futures, specifically their families. Although some of my own concerns form the backdrop of the writing, the book is not a personal reflection. Instead, it is a response to conversations I've had over and over again with students, friends, and acquaintances. My students are particularly concerned about this question: as people in their late teens to early twenties, they care about a range of issues, including climate change, and also their friends, families, and what they will do after college. They are particularly concerned with whether, and how, they will be able to build stable financial futures and have good lives, including families of their own.

But they are worried—*really, really* worried—about how the intensifying climate crisis is increasing their sense of uncertainty about all of those things. They're expressing that worry in classroom discussions, written assignments, and research projects. I've been seeing and hearing these discussions grow and intensify over the past few years, and I've been listening. This book is my response to their concerns and the concerns of so many other people in their reproductive years. It was demanding to be written.

While the questions at the heart of the book focus on the future (many of the people interviewed do not have children), they are also informed by the present: what we currently know about climate change, the ways we see it disrupting our own lives and those of others, and our expectations of how it will worsen. And despite the frequent lack of attention to race in these issues, these are racial questions. As my interviews and survey data demonstrate, racism is deeply intertwined with anxieties about the climate, the future, and the ability to raise children through it all. Like poverty, racism makes people vulnerable; it robs us of any sense of certainty that we will be able to secure, and maintain, the resources to ensure that we will not only survive but thrive. And in this case, racism makes climate-concerned people of color feel uncertain about their ability to have and raise children who will thrive in the midst of another overwhelming crisis.

To write this book, I drew on hundreds of published articles and books, interviews with a couple dozen Millennials and Gen Zers (and a few people in older generations too), and the results of a national survey I conducted in August 2021, which drew more than twenty-five hundred respondents between the ages of twenty-two and thirty-five. I also delved into how social media has become a vehicle for conversations about rethinking parenthood in the current moment. And, over and over, race and inequality arose as key issues shaping the discussion.

In a broader sense this book is a new way of responding to questions I've been asking for a long time about climate change and reproductive concerns. My first book, *On Infertile Ground: Population Control and Women's Rights in the Era of Climate Change*, tackled these issues through an analysis of how historical, scientific, political, and other developments have shaped our understanding of the relationship between reproduction and the environment, through a focus on population growth. That book was a critique of the history of population control as well as a deep dive into how inaccurate assumptions about population growth have been pervasive throughout ecological science, environmental activism, climate science, and everyday understandings of environmental problems and their solutions. It was a treetops-level discussion; now I'm turning to the ground-level concerns.

How do these big issues—environmental sustainability, climate change, population, the impacts of reproductive issues in general—land in individual people's lives? How do people think about their own reproductive concerns in relationship to larger social, political, and economic issues? In other words, what's the relationship between a big, collectively shared problem—climate change—and intimate concerns over pregnancy, birth, adoption, or choosing to be child-free?

These questions are complicated by the fact that, as I write this, multiple intersecting crises in the United States and beyond are heightening the stakes of the issues discussed throughout this book. Reproductive rights and bodily autonomy are threatened by the *Dobbs v. Jackson Women's Health Organization* ruling by the Supreme Court, which overturned *Roe v. Wade* and ended the Constitutionally protected right to abortion in June 2022. Racial violence against Black people continues at the hands of vigilantes and police, and racist ecofascists murder Black and Brown people in the name of so-called environmental sustainability. Climate chaos is all around; weird weather is not only a source of everyday conversation but the backdrop to community-level catastrophes and displacement, right here in the United States. Extreme political polarization and xenophobic discourse infuse how media outlets report the news. In the mainstream media, systems-level analysis of the climate crisis and its needed solutions are in short supply.

It's no wonder that young people are distressed and worried about the future and whether they can, or should, bring children into this

world. But I have also found that grappling with this very question—
whether, why, and how to have children, specifically in the midst of
the climate crisis—offers an important opportunity to rethink ethical engagements with ourselves and with the people and planet
around us. Specifically, it can inspire us to become more deeply
invested in creating the future we'd like to live in, whether we raise
children or not. This book offers a cautiously hopeful approach to
these questions.

Introduction

NO FUTURE FOR US

Victoria and I are talking via the computer on a spring day in Southern California.[1] She is a young Black woman from the Inland Empire region of California, where the diesel trucks on the road and intense heat collide to create some of the worst air quality in the country. At twenty-three years old, Victoria recently graduated from college and moved back home with her parents while she prepares to apply for graduate school. Although she is excited about the next stage of her life, she is also afraid—and those fears have a lot to do with climate change.

"There's a lot of fear and a lot of frustration about what's ahead of us with climate change," Victoria says. "This isn't normal. All of these brush fires, I'm like, the world shouldn't be catching on fire like this. Like, that's not normal, for the air to be smelling like smoke for months at a time. Like, I'm just seeing chaos. Things that are not what I would expect to see, like crime rates going up and environmental problems everywhere, but when I turn on the news, they're just brushing over it. But it's getting worse and worse. Why aren't they focusing on it?"

Victoria wants children—lots of children, she says, as many as four. Her parents immigrated to the United States from Ghana before she was born, and she was raised in a large, happy family of cousins and extended relatives. Although she wants the same for her own future, she doesn't believe it will happen, because of increasing environmental instability. She worries how that instability is impacting her.

> If there's a catastrophe like a hurricane, if I had kids in that situation, I would be worried every day about their life and them being alive and getting to grow up. Like, would I be able to keep them safe? That's a really scary part of parenthood that I don't have to think about now because I'm not a parent. It's frightening to see the world, how there's so much more chaos happening. And it's environmental chaos . . . like, do I want to bring kids into that if I don't even know what the future holds at all? And being a person of color and thinking about the future and having children . . . it always comes with some type of other narrative when it comes to our bodies and children. The way we're thought of and treated. Whenever I think about it, I just feel really anxious.

Anxiety. Worry. Fear. Frustration. Despair. These emotions and others have increasingly become the subject of surveys and other research into how climate change impacts emotions and mental health outcomes; for example, a 2020 poll conducted by the American Psychiatric Association found that 67 percent of people between the ages of eighteen and twenty-three were either "somewhat" or "very" concerned about how climate change is impacting their mental health.[2] These emotions also shape how people of reproductive age grapple with what I call the "kid question": the issue of whether, when, and how to have a child while grappling with climate crisis. A nonscientific national opinion poll conducted in 2021 found that 75 percent of members of Generation Z and 77 percent of Millennials said that climate change affects their major life decisions, including where they want to live, their career paths, and parenthood. On that last note, 78 percent of Gen Zers in the survey said that they weren't planning to have children because of climate change.[3] In other words, climate change is having a direct impact on plans for the future, and the kid question is a part of those plans.

The widespread impacts of climate change—a set of long-term shifts in temperature and weather patterns, unevenly spanning the globe since the 1800s as a result of human beings burning fossil fuels, cutting down forests, and using land in unsustainable ways—are disrupting the planet's ecological systems, which we depend on for survival. These disruptions include warming temperatures, droughts and water scarcity, heat waves, sea-level rise, increased and intensified storms, flooding, loss of biodiversity, and severe wildfires. Climate change is causing excess human deaths from heat events, malnutrition, malaria, and diarrhea, and is making us sicker from a range of health concerns from asthma to cardiovascular problems.[4]

Scientists concur that the changes to the climate are intensifying, and will worsen to the level of being catastrophic, unless government leaders around the world unite to take strong policy action to reverse reliance on fossil fuels and build energy infrastructure that centers alternative energy.[5] In other words, we're in the midst of a crisis. So far, progress toward these goals is limited and uneven on a global scale, and the United States—a global outlier in terms of the historical number of emissions the country has put into the atmosphere—has been notorious for refusing to act aggressively to enact necessary policies. Even when US government officials do adopt key legislation to combat climate change, the gains are inconsistent and uneven. For example, President Joe Biden signed the Inflation Reduction Act of 2022, aiming to reduce carbon emissions by approximately 40 percent by the year 2030. The bill went through a nasty political fight in Congress but was a significant step in the right direction—particularly because it pays attention to issues of inequity and justice. And then, early the following year, Biden's administration approved the Willow oil drilling project in Alaska, which will generate up to 9.2 million metric tons of carbon emissions every year.[6]

The pace of climate politicking and policy making is dizzying and confusing. Meanwhile, humans, animals, and the planet itself are suffering. We're not just suffering physically: researchers are finding that climate change is having impacts on our emotions and mental health. These impacts are driving people in their reproductive years to question whether it makes sense—physically, morally, and ethically—to birth and/or raise children. In other words, they are asking the kid question. Newspapers, magazines, blogs, podcasts, and academic research journals have been publishing articles, surveys, and simple opinion polls on climate change–related emotions and kid questions for years. However, there's been a strange omission in these writings: they almost always ignore the issue of race. This is strange because climate change in the United States has disproportionately hard impacts on communities of color.[7] The same communities often have fewer resources to adapt and be resilient to climate disasters. You would think that race and inequality would come up at some point in the research on the kid question.

They don't. Much of the research about climate emotions—a lot of it focused on "eco-anxiety"—focuses on the experiences of young, white, middle-class people. While eco-anxiety may seem like a neutral term, climate emotions expert Sarah Jaquette Ray argues that it is often deployed in ways that center white people's experiences and reinforce white privilege.[8] Specifically, she writes that many public accounts of eco-anxiety, or climate anxiety in particular, reflect racial anxieties in which white people are anxious to hold on to their way of life and to "get back to normal" rather than confront and fight for systemic change. These descriptions frame climate change as the greatest existential threat of our time, a perspective that ignores the long-standing existential threats that have faced communities of color—including slavery, colonization, and the ongoing terror of police violence. In other words, accounts of climate change that describe it as the greatest existential threat of our time are rooted in privilege, because they ignore the experiences of communities living through other more immediate or long-term threats to their existence.[9]

These issues came into sharp clarity one day in June 2022, when I met my friend Laurie at a café for lunch. It was nearly a hundred degrees again; the temperature had been in near the triple digits every day for weeks. Laurie and I hadn't seen each other in a few years. She'd had a baby in the interim, and I was curious to hear how life was going with her little one in the mix. But as she spoke, the conversation strayed unexpectedly from diapers and potty training to a subject closer to my own heart: climate change. "I think about it all the time," she said, taking a bite of her sandwich. "When I was pregnant, I actually started having nightmares about it. I would wake up in a cold sweat, thinking, how could I knowingly bring a child into this horrible climate situation? Was this the most selfish act in the world? How could I ever feel good about being a mother, knowing what would await my child?"

Laurie is white, middle class, and married. She has a solid career. She is someone whom you would think of as privileged; she has easy access to fresh, healthy food. The air she breathes, despite being in a city, is relatively clean. She is not what you would call vulnerable. And yet she sees climate change and the ecological upheavals it produces as an existential threat: the kind of challenge that upends the possibility of continuing to live life as we know it. When Laurie became pregnant, she was consumed with alternating emotions: excitement and anticipation at her baby's arrival, and anxiety and worry about climate change. She was grappling with the kid question.

Of course, many of us may experience heat waves, intense storms, drought, or floods at some point, depending on where we live. But we don't all experience these effects in the same way, and the resources we bring to the experience—as well as what we have available to cope with afterward—vary greatly. Climate mental health expert Britt Wray—who, like Laurie, is white and middle class and has confronted strong, painful emotions as she thought about whether to have a child during the climate crisis—argues that people with more social privilege are particularly ill-equipped to deal with existential threat, because they're used to thinking that they are generally safe and that their government and other leaders generally act to keep it that way. Yet, she argues, it is important not to individualize the experience but rather approach it from a political perspective. "We're going to get this wrong if we depoliticize this pain, by not seeing its entanglement with centuries of environmental violence, racism, and domination," she writes. "Without that context, we cannot be honest about who is the most vulnerable now and going forward, nor figure out how to best reduce harm. On the flip side, the tumultuous feelings that are on the rise are completely valid, need tending to, and present a great opportunity for justice-oriented personal, environmental, and social transformation."[10]

While women like Britt and Laurie are often the public face of this struggle, they are not the only ones grappling with the kid question in the midst of climate change. In fact, a national survey of forty-four hundred Americans conducted in 2020 found that, of those who did not have children, the people who identified climate change as a reason why they were not parents were disproportionately Hispanic/Latino or Black. Specifically, of those surveyed, 21 percent of white respondents indicated that climate change was a factor in why they did not have children, compared to 41 percent of Hispanic respondents and 30 percent of Black respondents.[11]

Clearly race is a factor in how climate change shapes the kid question, but how and why? What's race got to do with it? This book asks these questions, exploring why more and more people of reproductive age are thinking of foregoing having children, or worrying about how to raise the children they do have, because of the climate crisis—and what role race plays in these concerns. For example, a widely cited Yale survey conducted in 2019 found that Hispanics/Latinos and African Americans care more about climate change than their white counterparts.[12] Researchers in the study found that 69 percent of Hispanics/Latinos and 57 percent of African Americans reported that they were either "alarmed" or "concerned" about climate change, whereas 49 percent of white respondents were. In contrast, 27 percent of white respondents were "doubtful" or "dismissive" of climate change, while

these numbers were just 11 percent and 12 percent for Hispanics/
Latinos and African Americans, respectively. These feelings of concern
may also translate into mental health outcomes; another Yale study
conducted several years later found that Hispanic/Latino people were
more likely to report climate change-related anxiety, depression, and
psychological distress compared to Black or white populations.[13] That
groups of color are more concerned and alarmed about climate change
makes sense: Black communities are most likely to be located near oil
refineries; Latinos often live in food deserts—communities that have
little access to stores that provide fresh, healthy fruits and vegetables
at affordable prices; and Native American tribal lands are crisscrossed
with oil pipelines that violate access treaties and poison aquifers.

Communities of color are also disproportionately exposed to
long-term air pollution, made worse by climate change, which has
myriad health impacts, including on newborn infants and children.
Racial inequalities and redlining make it more likely for people of
color to live in communities that are disproportionately exposed to
environmental threats. It makes sense that this exposure translates
into heightened worry and concern. Do these worries and concerns
also shape reproductive questions for communities of color? How
can we know if researchers who study climate and reproduction
don't also study race?

At the same time, climate and reproduction are increasingly in
the public discussion because ideas about vulnerability are shifting
and expanding. We are all vulnerable to climate impacts, though not
equally—and many people who otherwise may have never thought
of themselves as vulnerable to existential threats are coming to grips
with the intensity of the crisis. Add to that the stunning blow deliv-
ered to reproductive autonomy in the June 2022 repeal of *Roe v.
Wade*, and eco-anxiety and the kid question are hot topics. But the
public attention they are given often ignores those who are most
deeply impacted—either through exclusion or through assuming
that young, middle-class white people's experiences and perspectives
are shared by all. They are not.

This book draws on dozens of interviews with people of color in their reproductive years as well as the results of a national survey I conducted with more than twenty-five hundred respondents. I also read and analyzed hundreds of articles, books, and various forms of media to explore how people are thinking, talking, and writing about how climate crisis impacts the kid question for them. Throughout, I demonstrate that these issues are simultaneously heightened by social inequality and that the challenges of struggling with the kid question are worsened because the struggles are largely waged in private. Yet these are not private concerns at all; they are the result of collective conditions—the impacts of climate change—that we are publicly living through every day as a society. Therefore, they must be addressed in public, in ways that prioritize climate and reproductive justice through systems and policy change.

The central arguments of the book are as follows. First, research and popular writing about climate concerns and reproductive anxieties tends to ignore race, with the effect of both erasing the experiences of people of color and low-income communities while representing white, middle-class perspectives as the norm. This matters because climate change impacts are distributed in much the same way racial inequalities are, and the impacts of those inequalities further disadvantage already marginalized communities, making it difficult to advocate for justice-oriented policy change. Second, climate-reproduction concerns are often responses to distressing environmental emotions, and when these emotions are relegated to the space of private concerns and individual decisions, it obscures the public nature of the problem. Climate change is a public problem that we are experiencing together, even if its impacts are unevenly distributed. And climate emotions are the result of this public problem. As a result, we are internalizing a set of concerns that we, as individuals, did not create—which makes it harder to demand the necessary action from lawmakers, corporations, and other powerful institutions that are causing us to feel this way. Furthermore, individualizing the problem and obscuring its racial dynamics allows dangerous groups like ecofascists to fill in the gaps by identifying population growth as the main problem—and racist violence as the horrifying solution.

The third and most important argument is that we're not listening to young people. Young Millennials and Gen Zers have been telling the people in their lives—each other, older relatives, teachers, and now the general public—that the climate crisis is disrupting their hopes, dreams, and imagined futures. For many, the climate crisis is instilling a sense of fear, making them question the moral and ethical consequences of creating families. For others, it is sparking new ways of thinking about how to build those families outside of a narrow biological model—from fostering and/or adopting to intentionally creating extended, blended families and kin networks. And for others still, they are foregoing raising children altogether.

There is no single way to respond to climate emotions, and there are certainly many ways to form a family if you desire that. But the problem is that increasingly people in their reproductive years aren't acting on their desires and dreams; they are foreclosing and narrowing their reproductive possibilities because the social and environmental conditions we're living in today are cause for despair, and they fear that the future will not be any better. Those I've interviewed for this book aren't giving up, though: some have organized public campaigns or nonprofit organizations to have the conversation with others in public. They've refused to focus on the narrow question of whether to parent children; they're instead claiming that the very fact that they have to ask this question is evidence of government policy failures. They are telling us that we need to change the fossil fuel–powered economic systems that drive the global climate crisis and make them feel that they can't have the families and futures they want. They are resisting these conditions, and they are asking us to join them. This book is my way of telling them: *Yes. I'm with you. And I hope this book will bring others along as well.*

What I don't discuss much in this book is population growth, at least not as a driver of climate disruption. As I detailed in my previous book, *On Infertile Ground: Population Control and Women's Rights in the Era of Climate Change,* population growth among the poor has been blamed for environmental problems for centuries.[14] This blame has been used to justify coercive interventions targeting communities of color around the world, resulting in sterilizations, abusive medical experiments, and other human rights violations. In the case of climate change, this blame particularly obscures the roles of governments, corporations, and the military—all of which are driving the problem, far more than the reproductive practices of people of color and low-income people. Focusing on population is dangerous. This is not a generalization; for decades, white supremacists known as ecofascists have seized on environmental issues to argue that low-income people of color are over-populating the Earth, and they have used these arguments to justify mass murder.[15] I will not give those ideas much of a platform here, in part because this book is about people who are primarily concerned about how planetary changes will harm their children, not the other way around. When I do address population and ecofascism in later chapters, it will be to critique and dismantle their underpinnings. All of which is to say, if you're looking for a book that will promote slowing population growth as a way to help mitigate climate change and save the planet, this is not the book for you.

Instead, this book is written from a place of advocating for climate justice—a perspective that seeks to redress the differential and unequal burdens of climate impacts on vulnerable communities—and reproductive justice. Reproductive justice comprehensively supports the ability to *have* the children you want, not have the children you *don't* want, and raise those you *do* have in safe and sustainable environments. Understanding climate anxiety and its relationship to reproductive anxiety and the kid question adds a dimension to the multifaceted fight for justice by revealing the ways the unjust systems causing climate change have pervaded not only our environments and communities but our minds, hearts, and deepest desires around children and family. That this is also a heavier burden for people of color is an injustice, and focusing on emotional and mental impacts adds an important dimension to the fight for justice on both the climate and reproductive fronts. With that said, let's learn more about climate-driven emotional distress, and how it is seeping into our most intimate life questions.

ECO-ANXIETY, REPRODUCTIVE ANXIETY, ALL THE ANXIETIES

I teach environmental studies classes at a university, and reproductive anxiety is a topic that comes up regularly among my students. Reproductive anxiety is closely related to eco-anxiety in general, climate anxiety in particular: it is a mental and emotional reaction to the moral, ethical, and social concerns one has about birthing and raising children during the climate crisis. My students are steeped in this anxiety, but they are not alone; I hear it from people I interact with at social gatherings, dinner parties, in online meetups, and on my podcast, *Climate Anxiety and the Kid Question.* If they're roughly between the ages of eighteen and forty, they're thinking about it, talking about it, worrying about it, talking to their friends, and feeling disconnected from and gaslit by older people who dismiss their concerns and tell them to have a kid anyway. It has taken root in their emotions, makes them worry about their mental health, and heightens a sense of feeling unsafe in the world—not because they aren't doing anything about climate change, but often because the volume of information they consume is overwhelming, and the actions they take to protect the environment don't seem to be able to make a difference on a large scale.

Daniel Sherrell's book, *Warmth: Coming of Age at the End of Our World,* captures the deep emotions that climate change arouses and the sense of isolation that can result: "The grief had its own weather. It would come down like a squall, momentary and encompassing, impervious to forecast. This was a weight I kept private mostly, unsure of whether or how to share it. Even with my closest friends, discussions of the Problem tended to stumble into the arid gully of knowing commiseration. 'We're so fucked' is what we let ourselves say, on the rare occasion the conversation wasn't quickly diverted into lighter terrain."[16] These are not private issues, but they are often assumed to be—which makes the discussion of climate-driven reproductive anxiety a stigmatized topic. When people feel that they can't discuss the issue publicly, like Sherrell, their distressing emotions are amplified through isolation.

Yet feelings like climate grief are common—so common, in fact, that researchers have been studying them for a couple of decades now. In the early 2000s, philosopher Glenn Albrecht wrote about the disrupted relationships between the physical body, the mind, and the external physical environment leading to a cycle of escalating stress, which can cause emotional and mental health outcomes to develop. He described a condition called solastalgia, the distressing sense of loss and alienation that arises when your sense of home is lost *while you still live there*.[17] This can lead to feeling a sense of nostalgia for a place that is literally still here, while simultaneously being lost due to climate change. Albrecht predicted that solastalgia and other psychoterratic conditions, or Earth-related mental health syndromes, would likely get worse as climate impacts worsen, particularly for more vulnerable people.[18]

Everyday climate emotions can lead to environmental melancholia, a form of mourning.[19] And ecological grief can result from losing beloved landscapes and cultural practices.[20] This grief does not have to arise from new experiences; for Indigenous peoples, in particular, climate change is a painful, ongoing reminder of the destruction of their ancestral lands, waterways, and air—destruction that began at first contact with colonizing Europeans. Part of this ecological grief is rooted in the feeling of loss of an anticipated future—and this is often expressed by people who struggle with reproductive anxiety. Naomi Klein described these feelings in-depth in her book *This Changes Everything: Climate vs. Capitalism*. While trying to get pregnant, and later, after having her son, her fear, worry, and grief were just as present as the feelings of excitement and anticipation at having a much-wanted child. As she thought about her child, how different his experiences in nature would be compared to hers, she was no longer able to enjoy her time outdoors because she was already grieving its inevitable loss. Imagining bodies of water covered irreparably in oil, or species of fish irretrievably lost to future generations, Klein experienced a sense of "pre-loss"—a morbid preoccupation with the inevitable losses that would result from climate change.[21]

Just having the knowledge that climate change is likely to continue—and worsen—can provoke emotional responses such as anxiety, grief, fear, despair, and even psychological distress and trauma.[22] In the context of an impending future full of uncertainty, the feeling can best be described as dread. David Theo Goldberg writes evocatively of this feeling, of how "dread drips from the pervasive sense that incomprehensible and indecipherable extinction or disappearance could strike in a blinding second."[23] However, dread is not experienced equally. Social marginalization—including racism, poverty, gender, sexual orientation, disability, and citizenship status—shapes the uneven ways people are exposed to environmental and climate vulnerability as well as the discriminatory provision of disaster relief services.[24] As a result, the ongoing legacies of colonization—of land, resources, cultures, and now the atmosphere—intensify not only the lived experience of climate change but also its mental and emotional effects. This phenomenon, which geographer Farhana Sultana refers to as climate coloniality, leads marginalized communities of color to feel a sense of alienation and rage, exhaustion and injustice, *in addition to* the broader climate emotions of mainstream society.[25]

That said, there is no single way to feel about climate change—people have diverse experiences, and diverse emotional responses. Scholars of climate emotions find that people express many emotions in response to climate change, including surprise, threat, sadness, strong anxiety, strong depression, guilt, and shame—and positive emotions too, such as motivation, joy, hope, optimism, belonging/connection, and love.[26] However, much of the research and popular writing in this arena reduces the complexity and diversity of these responses to one term: eco-anxiety. Although it is useful to have a shorthand way of referencing disturbing environmental emotions, the term can obscure more than it elucidates. Scholars and clinicians have been working to untangle these threads for some time.

Climate emotions researcher Panu Pihkala describes eco-anxiety as multifaceted, incorporating basic elements of anxiety such as uncertainty, unpredictability, and uncontrollability; it also includes existential anxiety and is often associated with feelings of overwhelm.[27] Climate anxiety is one aspect of this wider phenomenon but more narrowly focused on emotional and mental responses to anthropogenic climate change. Of course, the lines between eco-anxiety and climate anxiety often blur, because climate change is associated with so many ecological disruptions. But the point is that climate change requires a constant adjustment to changing circumstances in the midst of uncertainty, while also negotiating ambivalence about our ethical responsibilities as we attempt to keep a healthy perspective.[28] It requires an orientation to one's own ethics and visions of the future—a future that we expect to be characterized by ongoing disruptions of the present.

To be clear, eco-anxiety is not necessarily an anxiety disorder in the realm of a mental health condition but rather a moral emotion based on "accurate appraisal of the severity of the ecological crisis."[29] In some cases, people can experience a practical form of eco-anxiety that drives them to seek more information and change individual and collective behaviors for the better. Climate psychologist Caroline Hickman sees it as an "emotional gateway" into understanding more complex responses; it is a useful term that "allows us to speak with others of our fears of environmental and ecological vulnerability, collapse, and extinction."[30] In other words, it gives us a language for putting words to the feelings and having the conversation with others rather than suffering in silence. But in some instances, eco-anxiety is not just about temporary emotional states; it may describe long-term emotional changes that impact mental health. This shows up clearly in the documentary film *Katrina Babies*, which follows members of a Black community from New Orleans that was severely impacted by the storm and its aftermath. In one scene in particular, a young woman weeps when recounting the storm's impacts. When the filmmaker probes further, asking about her emotional response on camera, she says that she's never had anyone to talk to about her feelings before. Multiple people throughout the film share similar stories of not having an outlet for deeply distressing emotions that have lingered with them for more than a decade and disrupted their lives since the hurricane. This is shocking, given that there has been abundant research to demonstrate that children, their parents, and broader members of community in New Orleans and surrounding areas suffered ongoing emotional distress and more severe outcomes, including post-traumatic stress disorder (PTSD) and other mental health conditions, after the storm.[31] These outcomes were particularly high among Black New Orleanians.[32] And as climate impacts continue to grow in intensity and frequency around the world, so does a focus on climate-related mental health research.

CLIMATE CRISIS AND MENTAL HEALTH CRISIS

There are emotions, and there are mental health outcomes, and they are not the same. A fleeting sense of anxiety is an adaptive response to circumstances in a given moment; this is quite different from the kinds of chronic mental health disorders that interfere with daily life, cause inflammation in the body, and lead you to avoid potentially favorable interactions. Research shows that climate-related mental health challenges are growing—not just in the United States but around the world. For example, a comparative study of twenty-five countries found that negative climate-related emotions were associated with insomnia across all countries in the study; for those in Western countries, including the United States, these negative emotions were correlated with self-rated poor mental health, particularly among women.[33]

In a separate study, researchers looking at data representing more than ten thousand people from twenty-eight countries found that negative emotional responses to climate change, particularly climate anxiety, were negatively associated with mental well-being in all twenty-eight countries in the study.[34] Finally, in a third study, ten thousand young people between the ages of sixteen and twenty-five were surveyed in countries including Nigeria, the United States, France, Portugal, Brazil, and others; researchers found that the distressing emotions associated with climate change interfered with their daily functioning.[35] Survey respondents who lived in Global South countries such as Brazil, Nigeria, and the Philippines reported the strongest negative emotions in the survey—pointing to the fact that a larger sense of social, economic, and political vulnerability may play a role in climate emotions.

Some might be quick to dismiss distressing emotional and mental health responses to climate change as a form of giving in to doomism. Climate doomism suggests that we're past the point of no return: that there's little to nothing we can do to effectively mitigate the impacts of climate change, and that we should resign ourselves to our fate of eventual human extinction. To be clear, most people who are struggling with the kid question aren't doomers. While well meaning, the suggestion to counteract negative climate thoughts and concerns with hope and optimism doesn't actually work in this context. The kid question is an ongoing internal debate in which our individual actions, personal capabilities, and the resources available to us are situated in a broader context that largely feel outside of our control. It is a context in which presidents, Supreme Court justices, members of Congress, oil tycoons, corporate bosses, and the superrich play a huge role in putting tons of unnecessary greenhouse gases into the atmosphere, with serious long-term impacts for climate change.

And they aren't listening to the millions of people who are saying that they don't want this to happen. That it isn't fair. Whether you call it an emergency, a crisis, an urgency, or a situation, where we are with climate change is a hot mess, and a lot of people's emotions are a mess too. People are scared, anxious, angry. We don't know whether we'll be okay in the future, because we're not okay now. How could we not be anxious about that?

FORGET SOLUTIONISM

When I talk to people about my research, they usually ask one question: what's the solution? Well, I don't have one for you. It is not my goal to tell people what to do—certainly not to tell you whether to become a parent or not—but rather to suggest a new way of thinking, with the kid question at the center. I *am* here to tell you to reject the notion that distressing climate emotions are wrong, or that climate activism is a simple antidote that will replace climate anxiety with hope or optimism. Those so-called solutions ring false—and, as those I've interviewed have told me many times, trying to replace anxiety with action simply doesn't work. In fact, a 2023 report found that people who work in environmental organizations—on the frontlines of addressing our most pressing environmental issues—suffer intense burnout, including acute physical and mental pain, exhaustion, indifference, and hopelessness.[36]

Among them, Black respondents indicated that their burnout was caused or exacerbated by systemic and structural racism in the environmental and climate sector. Clearly there's ongoing repair work that needs to be done, far beyond simply "taking action" on climate change. Emotions are not problems to be solved. They shift, are amplified, become more complex or more clear, with the changing circumstances. Activists are among some of the most climate anxious people around, along with climate scientists. And, as climate psychologists reassure us, *your feelings are absolutely valid responses to the circumstances we're living in.*

That said, we can't give up. I, for one, am pretty attached to this planet we live on; I refuse to throw in the towel. And that attachment itself holds some of the seeds that can be used to deal with reproductive anxiety, grapple with the kid question, and make the families and communities we want. Hickman recommends reframing eco-anxiety as "eco-empathy" by focusing on the fact that people wouldn't feel eco-anxiety if they didn't also have a sense of caring and connection to people, other species, and the planet.[37] This kind of reframing allows for a discussion of connection, relationship, love, and understanding—and it helps to move away from the tendency to individualize and pathologize eco-anxiety and other distressing feelings.

There are other ways to reframe how our emotions are connected to the environment in this moment of climate crisis. In the book *Not Too Late: Changing the Climate Story from Despair to Possibility*, writer Rebecca Solnit argues for approaching this moment with a sense of hope—one that requires accepting despair as an emotion, while not remaining there. This entails a recognition of the possibility that the future is the result of what we make in the present, and to know that "joy can appear in the midst of crisis."[38] The edited volume includes a series of powerful essays harnessing dynamic emotions and everyday action to address the climate crisis from a place of power, even in the midst of despair. It encourages readers to take everyday actions to fight climate change as well as to continually ask "what can I do next?" As climate activist Mary Annaïse Heglar writes, "the right time to start your climate commitment is always right now."[39]

But that doesn't answer the kid question. What to do about that? First, we have to begin with recognizing that most framings of this issue shift the blame from corporations and governments onto individuals, thereby allowing those institutions to continue to be bad actors. Writer Meehan Crist argues that this blame-shifting accepts the business-as-usual of the neoliberal order—which is destroying the planet. This suggests that responses to climate crisis have to work within the established neoliberal order. This framing assumes that everyone, everywhere, pollutes in the same way, which we know is not true.[40]

We can resist those framings while recognizing that the kid question can operate as a springboard for recognizing and exploring deeper connections to care for planet Earth and each other. It helps us recognize the commitments we make to negative, fear-based mind-sets, so that we can release ourselves from those commitments. It helps to clarify what really matters and identify a sense of connection to something larger than yourself, including the possibility of leaving a legacy that will endure long after you leave this earth. Thelma Young Lutunatabua, a climate activist who decided to become pregnant while fighting the climate crisis, sees having a child as a radical act of hope that demands a commitment to a better future. She writes: "I completely honor and respect people's decisions to not have children because of climate fears. For me, having a child in 2022 is an act of radical hope. . . . As a mother, cynicism is not an option. My hope is tied to an existing practice of refusing to allow apocalyptic prophecies of the future to come to pass. It is a literal way of having more skin in the game. There is no choice for me but to ensure as many families as possible can safely make it through this rebirth."[41]

That legacy need not come through birthing or raising children, but this conversation can be a way in. It can help clarify what your environmental values are, whether and how you might get involved in climate activism, and how to begin to understand that the climate conditions we're living in, as well as the emotions they evoke, are neither individual nor private but rather social and collective—and they demand a social and collective response.

Ultimately what I'm saying is this: commit to fighting for this precious planet we live on. If the kid question gets you there, grab it and don't let it go. And together, let's build the futures and families we want, need, and deserve.

THE BOOK

Chapters 1 and 2 frame reproductive anxieties, and responses to them, within larger historical and political contexts. Chapter 1 offers a historical analysis of the diverse ways women and parents in the United States have enacted stances of reproductive refusal, resistance, or resilience. The chapter illuminates how we can become both reproductively and emotionally resilient, whether we are parents or not. Chapter 2 follows with a structural analysis of how climate change offers new opportunities for more people to align with the struggle for reproductive justice, particularly through advocating for the ability to raise children in safe and sustainable environments. It critiques the ways reproductive issues are privatized even as collective vulnerabilities worsen and calls for more expansive advocacy against reproductive oppression.

Chapters 3 and 4 situate the kid question within the reproductive anxieties of people of color in their reproductive years. Chapter 3 presents the results of a national survey I conducted on race, reproduction, and emotions in the United States. It also calls on other researchers to prioritize marginalized populations in their work as a matter of social justice. Chapter 4 shares the results of my qualitative interviews with a diverse group of people of color, both Millennials and Generation Z, presenting their firsthand accounts of how climate emotions impact their reproductive plans.

Chapters 5 and 6 follow climate activists as they create advocacy campaigns that address reproductive anxiety. Chapter 5 explores the race and gender dynamics of Conceivable Future, an organization that creates space for exploring and addressing climate-and-reproductive anxieties in private and in public. Chapter 6 looks at two large-scale public campaigns, BirthStrike and #NoFutureNoChildren, and tracks the ways antinatalists and ecofascists thwart progressive public discussions of reproduction, including those that center calls for justice and systems change.

Finally, the conclusion brings us back around to climate emotions, landing on what I think of as the most important one of all: love. Love of the planet, love of family and children, love of art and activism—the conclusion argues that a deeply held, passionate commitment to both human communities and the nonhuman world around us is necessary in this moment, and that it can help us answer the kid question in unexpected ways.

1 Childfree Nation?

In July 2019, singer-songwriter Miley Cyrus gave an interview to *Elle* magazine in which she discussed her thoughts about the environment and why she doesn't have kids. "We're getting handed a piece-of-shit planet, and I refuse to hand that down to my child," she said. "Until I feel like my kid would live on an Earth with fish in the water, I'm not bringing in another person to deal with that."[1] Referring to her generation of Millennials, she added, "We don't want to reproduce because we know that the Earth can't handle it." Cyrus wasn't speaking in generalities. Months before the interview was published, the home she shared with her then-partner Liam Hemsworth burned to the ground in the Woolsey Fire, one of several that ignited on the same day in November 2018. By the time it ended two weeks later, the Woolsey Fire had burned close to 100,000 acres, killing 3 people and prompting 295,000 to be evacuated from their homes.

Cyrus was talking about the kid question, and a lot of Millennials and Gen Zers are making similar statements in newspaper articles, on social media, and on podcasts. Despite their bold declarations about not intending to have children, however, they aren't actually talking about not *wanting* to have kids. Rather, they are identifying that the conditions of life on a climate-disrupted Earth are not ideal for having babies or raising children. That is not a personal decision; there's nothing private or individual about the moral, emotional, and psychological struggle facing those who might want to have children but feel that they can't because of climate change. Instead, this is a response to a public, social problem that we are collectively experiencing. And when people like Cyrus make these kinds of statements in public, they are engaging in what I call "reproductive resistance."

Reproductive resistance is an active stance that critiques and fights back against the unreasonable social, political, economic, and environmental conditions potential parents face during their reproductive years. On the one hand, reproductive resisters may publicly declare an unwillingness to have children in the current moment due to a sense of outrage at the state of the planet; or they may make their plans quietly, attaching reproductive decisions to the broader policy or environmental context. On the other hand, reproductive resisters may want to have children, and in some instances they do birth or adopt them, while still critiquing the lack of supportive conditions that make it possible for parents to feel confident raising children in this society.

This is quite different from what I call "reproductive refusal." Reproductive refusal is an active and ongoing set of decisions to not birth or raise children. People who are childfree by choice are in this category. Those adopting this lifestyle have grown significantly in number since the 1970s, when policy shifts granted women unprecedented access to legal, safe contraceptives and abortion on a national scale. The Supreme Court ruling in the 2022 case *Dobbs v. Jackson Women's Health Organization* (which rescinded the Constitutional right to access abortion) has created significant barriers to practicing reproductive refusal in the current moment, however, a growing cultural movement celebrating childfree lifestyles makes it more likely than ever that people in their reproductive years will continue to find ways to not have children, intentionally and happily. I explore reproductive refusal, including the childfree-by-choice movement, to delineate how it is different from climate-driven reproductive anxiety.

Finally, this chapter discusses the "reproductive resilience" of parents who make do and raise their children in the midst of very challenging circumstances. Drawing from insights in psychology and social movements, I offer suggestions for how to deal with the kid question by facing climate change head on—specifically, by having same the kinds of difficult conversations with kids that parents from marginalized groups have had with their children for many generations.

RESISTING, REFUSING, AND BEING RESILIENT

The year of Cyrus's interview and in the years that followed, major newspapers began publishing articles featuring the voices of young women who argued that that the climate crisis makes it impossible to have children. In one such article in *The Guardian*, a twenty-five-year-old former nanny described how she had always planned to have children until one of the children she cared for developed pollution-driven allergies. After reading scientific reports about declining insect numbers, she felt that her decision had been made: environmental problems were preventing her from being able to have children. She stated, "We are hurtling towards disaster, and if I can bring awareness to the situation by sharing this personal choice that I've made, I'm willing to do it. It is sort of desperation. I don't know if I'll regret being so public about it one day." In the same article a twenty-three-year-old woman described the future as desolate, saying that having children has to be understood in that context: "That's why I am not having one—because I feel so desperately that it would be bringing a life into a future that does seem ever more desolate. Every time a friend tells me they're pregnant, or planning on having children, I have to bite my tongue."[2]

In a separate piece published two years later, a thirty-nine-year-old English teacher in the United States stated bluntly, "I refuse to bring children into the burning hellscape we call a planet." Noting her long-standing uncertainty about having children, she concluded, "Now, as I look at the state of the economy, shoddy global healthcare and climate change, I just feel like all my trepidation was well justified."[3] On the face of it, these may seem like statements from people who are child-free—those who have made intentional and informed decisions not to birth or raise children, based on a personal calculus of the costs and benefits of doing so. However, being childfree stems from a different mind-set than those of Cyrus and her cohort. In many instances it stems from an enthusiastic embrace of intentional nonparenthood. As a growing number of podcasts, blogs, books, and social media accounts demonstrate, there are many joys associated with not having children. These sites depict young women frolicking on vacation, happily refusing to conform to societal norms and family expectations, and declaring the power of living autonomously from a place of true reproductive freedom—which, for them, means the freedom not to reproduce.

The childfree lifestyle is a form of reproductive refusal: a clear declaration that they will not have a child, now or in the future, and is generally a voluntary choice. This is, of course, quite different from being childless—the common umbrella term for adult nonparents who may have many reasons for not having children, including financial circumstances, infertility and health issues, or not having the ideal partner to raise children with. The term "childless" is controversial, as it suggests that there is something fundamentally lacking in the lives of nonparents, but it is useful to note that many adults who do not have children are not making intentional, long-term, or permanent decisions about reproduction. They simply don't have a child or children, and the reasons may be myriad and complex, but for whatever reason, their nonparenting is not a refusal to reproduce. They are not childfree by choice.

Neither are people like Miley Cyrus or the young women profiled in *The Guardian*. Making a decision to forego having children because you're afraid that the Earth isn't a safe and sustainable place to raise them is not an affirmative, autonomous, or individual choice. Rather, it is the consequence of decisions made outside of your control by powerful actors; it is the result of ongoing harm to the Earth and its atmosphere, and a recognition of the emotional and psychological damage this causes humans as a result—a collective impact felt by many people who may otherwise want to raise families. It is an act of resistance against those impacts. Although the climate crisis is a relatively new reason for doing so, there is a long tradition of such resistance, the history of which informs how we think about children, the environment, and existential problems today.

WE HAVE ALWAYS RESISTED

Reproductive resistance—an active way of fighting back against the undesirable conditions shaping pregnancy, birth, and parenting—has been a part of the landscape in the United States since the country's origins. For example, Black enslaved women, whose bodies were used to breed children as labor for plantations, used various means to resist their circumstances. They used a range of tactics to avoid rape and forced marriage (common sources of pregnancy), chewed cotton roots for their contraceptive properties, used herbal abortifacients, and in some instances committed infanticide to prevent their children from growing up in the horrific conditions of slavery.[4]

More recently, reproductive rights activists fought throughout the twentieth century for access to safe contraceptives and abortion, both through advocacy for legal access to these tools as well as through calling for research to develop safe contraceptives. Even clandestine forms of resistance played out in public. For example, in Chicago an underground network of informal abortion providers called the Jane Collective hid in plain sight and provided women with sliding-scale abortions, performing approximately eleven thousand of the illegal procedures between 1967 and 1973.[5] This history is connected to the public work of Margaret Sanger, the well-known reproductive resister who opened the first birth control clinic in the United States in 1916. Sanger fought the Comstock laws, which made it illegal to send contraceptive materials—at the time considered to be lewd and lascivious—through the mail. She also advocated for research and development of new contraceptive methods that women could control for themselves. At the same time, Sanger was a highly controversial figure; she supported the discredited science of eugenics—which says that social traits and characteristics are inherited through DNA and can be controlled through selective breeding—and created the Negro Project, which promoted selective breeding to prevent so-called undesirables in Black communities from reproducing.[6]

Reproductive resistance has long been a part of the American environmental landscape too. Feminists began to align their efforts with those of environmentalists in the 1960s, when groups like Zero Population Growth worked with the National Organization for Women to advocate for birth control and abortion access. Although environmentalists' efforts were focused on population control, they also partnered with women's rights advocates to promote the idea—deemed radical at the time—of separating sex from reproduction and womanhood from motherhood, arguing that being able to plan their families (or opt out of having families altogether) offered women the opportunity to do other things in life, including pursuing education and careers. These efforts were explicitly concerned with existential questions; even though they weren't yet focused on climate change, environmental activists were also advocating for food security, clean air and water, and sustainable land use. Like others over the centuries, they publicly connected reproductive decisions to the broader circumstances that shaped them and actively resisted bringing children into a troubled world.

SO . . . ARE WE STILL HAVING KIDS?

While reproductive resistance has a long history in the United States, reproductive refusal is a more recent phenomenon, largely facilitated by policies expanding access to contraception and abortion. Beginning in the early 1970s, after the legalization of contraceptives and abortion access on a national scale, large numbers of American women began doing something women in generations past could not have imagined: safely and intentionally not birthing children. Today, close to one in five American women ends her childbearing years without giving birth. This number has risen for all racial and ethnic groups; although white women are most likely to never become biological mothers, the rates of foregoing children have increased for Black, Hispanic, and Asian women over time. In 2008 there were 1.9 million nonparent women in the United States, compared with 580,000 in 1976.[7]

In a 2021 Pew survey, among all nonparents between the ages of eighteen and forty-nine, 44 percent said that it is either not too likely or not likely at all that they will ever have children. While a small number (5 percent) cited environmental reasons, the majority said that it is because they simply don't want children. People under the age of forty in the survey were more likely to give that response (60 percent) compared with those forty to forty-nine (46 percent).[8] Adoptions are also on the decline; the number of children in foster care awaiting adoption has grown steadily since 2012, while adoptions of all kinds, which had been slowing steadily since 2007, dropped steeply during the Covid-19 pandemic.[9]

One of the reasons nonparenting has been on the rise is because of a recognition that raising children doesn't necessarily make people happy. Nevertheless, we are told, over and over, that having children is necessary for a happy life. This messaging is heavily gendered: women are frequently pressured by family and friends into having kids, or justifying their reasons for not having them.[10] The pressure isn't neutral either; it is attached to negative moral judgments about people who don't have children as well as assumptions that not having children is a personal deficiency.[11] Research conducted over more than three decades has consistently shown that nonparents are viewed less favorably than parents, particularly those who are child-free by choice; they are often perceived as less fulfilled and psychologically maladjusted. Surprisingly, these findings still hold today: researchers found in a 2017 study that men and women who are childfree by choice were perceived as leading less fulfilling lives compared to parents, and that their decisions to not have children evoked moral outrage in the form of anger, distrust, and disapproval. Rather than being a matter of personal choice, parenthood is still widely perceived as a moral obligation, and those who don't meet this obligation suffer the stigma of other people's negative judgments.[12]

Ironically, research conducted over the same decades suggests that many parents, particularly in the United States, are less happy than childfree people. A comparative study of countries in the Organization for Economic Co-Operation and Development (OECD) found that the parenting "happiness gap" is most pronounced in the United States.[13] This is largely because—despite being a society that stigmatizes nonparents—the United States does not offer parents the kinds of policies that would help them thrive while working and also having a balanced life. In the European countries where parents reported higher happiness levels, more generous family policies, particularly paid time off and childcare subsidies, are commonplace.[14] Such policies are important; so is everyday pleasure and enjoyment, which can be lacking for parents. In a 2004 study of 909 working women in Texas, researchers found that when asked to rank how much pleasure they gained when they engaged in a list of nineteen activities (including cooking, shopping, watching television, or doing housework), most ranked childcare sixteenth, *after housework*.[15]

With that said, parenting is in fact a source of happiness for a lot of people. A study of 1.8 million parents and nonparents worldwide assessed how they evaluated their own sense of well-being and found that parents with children at home experienced both more daily stress *and* more daily joy compared to nonparents. More important, the study found no significant differences in happiness and well-being between those who *chose* to parent and those who chose not to, while those who became parents and felt that it *wasn't* their choice were significantly less happy than others in the study.[16] This clearly points to the importance of being able to make autonomous decisions about reproduction, something that is significantly more difficult for many people following the Supreme Court's *Dobbs* ruling. It also highlights the fact that when considering reproductive "choice," what we're really talking about are the enabling or constraining conditions—including laws governing contraceptive and abortion access, financial and workplace policies, and social and cultural expectations—that make parenting feel like a desirable option or not.

WHO WANTS KIDS ANYWAY?

A young woman's social media account shows a video of her reclining on a boat, smiling and sipping champagne, with a stunning sunset in the background. It is a scene of pure relaxation and bliss, the kind of easy vacation lifestyle anyone stuck at home would envy her for. Other short videos paint a similar picture: she strolls through cobblestoned streets, sips coffee at quaint cafes, and snuggles under her blanket on airplanes. However, the scenes of bliss are always interrupted by a jarring sound: babies and children. Babies crying. Toddlers throwing tantrums. Children arguing with their parents over their nap time. Kids wail, parents try to soothe, and those around them sigh, some in sympathy, some in annoyance. In the captions she tells us that there is a reason, one reason alone, why she and her spouse are able to live their seemingly carefree lifestyle: because they don't have children. They are childfree, and very happy about it. Don't believe them? You don't have to. But while you're sitting there,

getting ready to change another dirty diaper or steeling yourself
against the nightly bedtime ritual/battle, she's going to be pouring
up another glass of champagne . . .

There is a difference between not having children and being child-free; that difference is in the word "free," with its connotations of an active, ongoing decision to not become pregnant, foster or adopt, or otherwise become a parent. In the decades in which women have had fewer children, a number of groups have sprung up to support and celebrate this decision. There is the NotMom summit, a conference designed to "connect, inform, and empower the growing audience of childless and childfree women."[17] On Meetup.com, a social media platform that hosts in-person gatherings and online discussions among people who share similar interests, as of early 2023 there were more than fifteen childfree groups with a total membership of just under four thousand. Designed to bring together women who don't have children, with names like No Kidding!, Ladies Without Little Ones, I'm Not Kidding, and Ladies With No Babies, the groups span cities across Australia, Canada, Europe, New Zealand, and United States.

For those who prefer to form their communities online, there are dozens, if not hundreds, of social media communities, podcasts, virtual meetup groups, books, and influencers who extol the virtues of the childfree lifestyle. The social media influencer Rachel Cargle, for example, proudly declares herself the "rich auntie" who enjoys being involved in the lives of her friends', family members', and partners' children while remaining childfree herself. There are culture-specific groups like No Bibs Burps or Bottles, a Black women's online community that describes itself as a group for the "chocolate and childfree"; the organization's leader offers one-on-one coaching, a virtual support group, and a podcast for Black women who want to hear culturally relatable content. Another podcast, the Berlin-based We Are Childfree, hosts women of all ages from around the world to have in-depth discussions on why they've chosen to forego having children. From careers, to travel, to simply not having the "maternal gene," they normalize the everyday reality of making ongoing decisions to prioritize aspects of their identity that don't include being mothers.

As noted, being childfree by choice is an intentional act of reproductive refusal, a way of saying no that is grounded in asserting one's reproductive autonomy. Yet the conditions of that autonomy are not individual or private, but rather social and political. In her writings about reproductive refusal, feminist scholar Carolyn Morell notes that reproductive "choices" and "freedoms" for women are socially determined, therefore we cannot address them without first dealing with the social, cultural, and political contexts that shape how childbearing decisions are made. These social and political contexts are based in race, class, citizenship, age, geographical location, and other aspects of identity. Furthermore, there is a lot of diversity in how people's specific experiences impact how they feel about parenting. For some, motherhood is *the* path to a meaningful life, whereas for others, saying no to parenthood opens up the possibility for a life of rich fulfillment. Having a language to destigmatize reproductive refusal and childfree lifestyles, and making it possible to talk about this path as a viable option, can help bolster reproductive autonomy for all women.[18]

While being childfree is often presented online as a personal lifestyle, it can also be thought of as a social movement. Sociologist Amy Blackstone, a researcher of childfree women, argues that it is a "grassroots effort to educate the public about nonparenthood as a legitimate life choice, raise awareness about the problems associated with overpopulation, and advocate for those who made the choice to be childfree."[19] (Note: Not all childfree people support these ideas about population; in fact, many Millennial and Gen Z childfree people reject the overpopulation argument.) The movement challenges assumptions that those who don't have children lack purpose in life and repositions parenting as an option rather than a requirement.

Although more and more young people are actively deciding against having children, as Blackstone notes, being childfree is nothing new. In fact, it moved into the public sphere as a formal social movement in the early 1970s, most notably with the founding of the National Organization for Non-Parents. Ellen Peck, the organization's cofounder, wrote the controversial 1971 book *The Baby Trap*, which outlined the movement in detail. Peck held troublesome views that were common to many liberal feminists at the time; she made antipoor and antiwelfare statements, suggesting that poor women had children in order to receive government benefits.[20] Nevertheless, childfree individuals and groups today are quite diverse in terms of race, class, and their support of women's ability to make a range of choices about childbearing.

Lesbian, bisexual, and queer women are often left out of the discussion, but they experience many of the same complex questions and struggles that heterosexual women do. Research shows that straight family members often assume that lesbian and queer women will be reproductive refusers, opting out of motherhood. However, this is not necessarily the case; many lesbians do want children—and in fact, many younger lesbians *expect* to become parents at some point in their lifetime.[21] Unlike heterosexual couples, many lesbians and other queer women enter their relationships expecting to form families in diverse, more-than-biological ways, including through in vitro fertilization (IVF), fostering, adoption, and step-parenting.[22] A comparative study of heterosexual, lesbian, bisexual, and queer childfree women found that regardless of sexual identity, there were two common experiences that were central to women's experiences: first, that being childfree is not a singular, permanent decision that takes place at one time, but rather a series of smaller, ongoing decisions that continue throughout one's reproductive years; and second, that their biggest struggle was to simply normalize for *others* that not having children was an ordinary decision they were making, rather than a radical declaration against motherhood.[23]

As noted earlier, the contexts of reproductive resistance and refusal arise from the broader social, political, economic, and environmental circumstances of people's lives. When those circumstances are

complicated, such as when potential parents face significant challenges to their health, livelihoods, or emotional and mental states,
some people have persevered and become parents while navigating
deep uncertainty and troubling circumstances. They have engaged in
what I call reproductive resilience, and their example points the way
to how people will increasingly have to navigate parenting in the climate future. Reproductive resilience, first and foremost, begins with
a recognition that the conditions for raising children may include significant challenges such as racial inequality, poverty, and unequal
gendered labor expectations. It also involves a commitment to birthing and/or raising children while navigating those challenges, particularly large-scale structural challenges, whether anticipated or not,
and summoning all of the resources at one's disposal to do so.

IT'S NEVER BEEN A GREAT TIME TO HAVE KIDS

When I began conducting interviews for this book, I was surprised
to discover that my own parents had wrestled with the kid question
in the context of nuclear war. From the time they married in the late
1960s until they had their first child in the mid-1970s, international
missile skirmishes and a protracted Vietnam War led them to have
multiple conversations about whether having children was a good
idea, particularly when they were worried that my father would be
drafted to fight in Vietnam. Like many climate-concerned people
today, my parents discussed the issue with family and friends,
watched disturbing images on television, and tried to manage their
reproductive anxieties. My father recounted these concerns to me:

> ME: When you and Mom were thinking about whether to have
> children, were you concerned about nuclear war?
>
> DAD: Yes, that was a definite concern for us. Were we all going to get
> blown up? We decided to just deal with it. You know, this
> thing about having children in the nuclear age was a personal

thing. We talked with our friends about it. Some agreed with us. Others said, "There won't be any nuclear war. God will protect us." We thought, if we stop having babies, there won't be any more of us in the future. I said, "Look, day-to-day life goes on and on," despite my panicked thoughts, which were overwhelming at the time. So we decided, we're going to have children like everyone else. We're going to pretend it's not going to happen. God will have to protect us.

ME: But was nuclear war actually something that seemed real to you?

DAD: Yes, absolutely. We got *that close* to a nuclear war happening. My brother was in the army, and my mother was walking around the house, freaking out, "They're going to kill my baby!" It was very real. With the Bay of Pigs, they came that close to pushing the button. Two white men sitting up there deciding the fate of the whole planet. It was just terrible.

These were not uncommon thoughts in the mid-to-late twentieth century. Britt Wray, in her book *Generation Dread*, writes that her parents held an "End of the World" party when she was a baby, fully convinced that a nuclear attack would soon end all life on Earth.[24] When Wray reached her reproductive years several decades later, she began to struggle with her own existential fears and the kid question, although her focus was on climate change. While Wray's and my parents lived in different countries and navigated different cultural and racial contexts, they were deeply immersed in the reproductive anxieties of their time, shaped by large-scale social, political, and economic circumstances, much like climate change today.

A brief, personal look at history demonstrates that hard times and life-threatening circumstances don't limit themselves to specific time periods; children live through existential threats all the time. As a schoolchild in the 1980s, I remember vague references to the Cold War rivalry between America and the USSR in the news and on television. The existence of nuclear weapons informed those messages, although it was unclear what they were and what kind of danger they posed. They were distant concerns; we were more concerned with the occasional earthquake or with our practiced fire drills requiring us to line up and march in formation to a designated safe space outside. Thirty years earlier, when my mother was in elementary school in the 1950s, she and her peers participated in air raid drills. Like my friends and I, they held hands, lined up, and followed their teacher to safety, in this case down in the school's boiler room, to wait out a potential bombing from the sky.

At the other extreme, today's school children participate in active shooter drills, learning what to do if a person with a gun enters their school or classroom and opens fire. Unlike my mother's air raid drills, active shooter drills are not hypothetical threats but rather an all-too-common literal threat to children's lives. Collectively, we've always faced the possibility of life-threatening events, and children have always had to navigate this danger too. As people have faced the possibility of having children, existential threats have long shaped their reproductive decisions and plans.

REPRODUCTIVE RESILIENCE

Reproductive resilience has long been the reality for so many families that navigate racism, poverty, homophobia, and other forms of oppression and marginalization on a daily basis. Reproductive resilience refers to everyday acts that build the tools, strategies, and community support necessary to survive and thrive in the midst of deeply challenging—seemingly impossible—circumstances. Women of color are called upon to do this all the time: children of color in the United States experience racist profiling and discrimination in schools, health-care settings, and social encounters from an early age; communities of color are chronically exposed to toxic air pollution as a result of living near oil refineries and landfills; given these kinds of cumulative exposures, zip code is a key predictor of health and well-being in the United States.[25] Taken together, these conditions are life-threatening; they literally pose an existential threat. As a result, mothers of color have long played a key role in activism for reforms in neighborhood policing and criminal justice, education, health care, and environmental justice.

This kind of reproductive resilience may be less familiar for white parents, but as climate change worsens, parents across race and class divides will have to navigate reproductive resilience in new ways and will have to do so intentionally. To find out more about reproductive resilience in the climate change context, I interviewed a group of mothers (and one woman who is trying to become pregnant) who ranged in age from their twenties to their forties, all parents of children under the age of ten. All of them were struggling with the same version of the kid question: How do you manage eco-anxiety as a parent? What do you do with the knowledge that the environmental problems facing your children will likely worsen over their lifetimes? Can you forge a parenting style that can ultimately serve as one of the legacies you leave for those who come after you?

Sarah, a white woman in her forties and a professor of environmental studies, said that parenting through climate change made her more aware of her privilege. "As you start to raise the question, I can feel my heart rate picking up, so my, I can tell you right now, just my body is responding to that. The concern I have there in part has to do with things that I realize are a function of my relative privilege, mine and my kids'. And that is that I had this idea because of my privilege, that any reproduction I did would involve my children having maybe a better life than I did, or for at least, you know, the pleasures of living on this planet. And that seems all of a sudden maybe not so true. It feels destabilized." The privilege Sarah described is both a protection from harm as well as an ability to experience the things in nature that she experienced when she was young. "You know, when I grew up myself, I had a sense that going out into nature, so to speak, was supposed to be a lovely healing thing. And sometimes I have access to that. But I don't think that my children, really frankly, even my own generation anymore . . . anyway, we'll ever have that kind of untainted view of nature. And this sense of the very literal ground that we stand on not being stable, is just the way my kids are always going to feel about the environment."

Carla, a Mexican American woman in her early twenties, expressed a sense of concern about the important cultural experiences and family traditions she and her partner grew up with that won't be passed to their child. "My partner spear fishes," she said. "He's been doing this, it's been in his family for generations and it's something we were hoping to pass on to our child. And I say to him, 'They're saying that by 2050 there will be more trash in the ocean than fish,' or I'll talk to him about dying coral reefs. These issues are a concern for me. I've noticed in the area where I grew up, we had birds singing right outside the windows all the time when I was a child. They're not there anymore. I read more and more about toxic pollutants in the air, or in the chemicals that are used to grow our food, and I'm concerned. I'm very concerned for my child. I don't want my child to grow up with that." In the midst of her concerns, Carla felt positive about her own family's future, though thinking on a global scale felt more distressing. "If I'm imagining the future on a personal level, it's hopeful. I'm excited. Um, happy. Yeah. I think those three words really encompass the feeling I'm looking forward to, new and exciting experiences. Inspirational. But if I'm thinking on a much larger level, on a global scale, then it's anxiety and fear and sadness."

While Carla was involved with campus activism in her student years, her preoccupation with climate change—and her status as a new parent—have recently inspired her to redouble her efforts, with a specific focus on anticapitalism activism. "Some people say there's no way. No one will ever want that. If that's the case, I don't know what we're doing here. I think the point of being alive on the planet as human beings is to try to make things better. To try to come up with solutions. Better means equity, consideration of everyone involved. Making life as livable as possible. Thinking and behaving

sustainably. If we don't think about and plan with social and political and environmental issues in mind, it won't be livable for long."

Britt, a thirty-five-year-old white woman who is a writer and edu-
cator, is also a new mother. She has written a book about climate
emotions, which she initially thought was spurred by her feelings
about her desire to have a child, but writing the book inspired Britt
to reflect further and realize the long-term nature of her environmental feelings. "I think there was always this emotional undercur-
rent about the damage that humans are causing, and the frustration
with my lack of ability to really reconcile that problem even within
my own life completely. So I think it started there. But then, later,
when it comes to this question of whether or not to have a child, it
became acute in a new way. The distress was both acute and chronic.
So it became stronger and fiercer, and it was very much long lasting.
It was staying with me, and changing the way that I felt in everyday
scenarios. And I think that it was because the existential stakes are
so pronounced on the question of whether or not you are going to
willingly bring a child into this world and commit them to a long life,
hopefully a long life of, you know, contending with a warming world
and all of the other intersecting dangers that are at play now."
Parenting became Britt's motivating force to address her climate
emotions directly and tackle her fear of the future head on. To her
surprise, pushing aside her ambivalence about having a baby
entailed a commitment to rejecting fear in favor of joy. "There were
many times, thousands of times, I thought I would not make the
decision to have a child . . . and then, finding myself, of course, coming to a point where we do decide to have a child, a lot of that is
about a commitment to joy. A lot of that is about producing the
future that I want to be living in."

Lily, a Black woman in her early forties, finds one of the biggest
challenges to feeling secure as a parent is the lack of reliable information, education, and policies to address climate change—and
she's clear that this is a government failure, not a failure of parenting. "It's hard to even have a place to go, to have information you can

trust," she noted. "Everything feels like a circus right now. Obviously climate change is real. We need better information about it and we need better, stronger government actions too. Things like regulating vehicle emissions or how our foods can be packaged, or buying single use consumer products. We need a much larger scale of like compostable spoons and things. Everyone was talking about Covid when it first happened, but this is way bigger than that and we don't discuss it at anywhere near the appropriate scale." Her fears intertwine with her desire to take action, but she often feels uncertain of how to move forward. "So I'm like, okay, we kind of know about it, we're all kind of scared about it, but what are we really doing? I don't think people really know what to do. And so, we need people in charge taking it really seriously and doing something. Putting policies in place and giving us information, what we can do at a local level."

Some of the mothers I spoke to had already been through their own personal versions of hell—existential threat, not hypothetical but real, up close and personal—and came through on the other side. One of them was Lydia, a high-school teacher in Paradise, California, who openly wept during our conversation as she spoke about the ongoing mental health challenges of dealing with a disaster's aftermath. The town of Paradise burned to the ground in the 2018 Camp Fire. "I've been struggling with really bad depression the past few months. A lot of it is environmental. For one thing, it's been smoky for a month where I live. *For a month.* And it's been like this the last three or four years, especially the last two, which have been really bad. And then you add the disease (Covid) to it, and it's not safe inside and it's not safe outside. And I just feel like there's no refuge that I can provide to my growing child."

Lydia is white and in her late twenties; although she had been sensitive to environmental issues for most of her life, she was optimistic prior to having her daughter, which changed to a sense of fear after Donald Trump was elected president and the general political and social climate became more deeply polarized. The Paradise fire compounded her feelings significantly. She offset her fears through an intentional approach to teaching at the high school. Lydia's students, many of whose families lost everything they owned during the Camp Fire, continue to struggle with compounding traumas from the disaster. As a result, Lydia brings a hopeful, optimistic attitude to the classroom. "I don't preach doom and gloom; they're young, and they've already been through so much. When I first started teaching, my expectations were academic. I wanted to prepare them for college. But now, post-fire, it's all about relationships. Relationships were important before, but now it's a very relationships-centered classroom. I meet them at the door, I learn their names immediately, I check in with them, I learn what they're about, I learn their interests and follow up. I have them pull out a sheet of paper and write five things they're grateful for."

Monica, like Lydia, has also lived through a disaster: war. Her family came to the United States as refugees when she was a child, escaping the Bosnian War. Now in her late thirties, Monica and her husband are actively trying to get pregnant. Her husband is worried about climate change; he simply doesn't understand Monica's strong desire to be a mother, which she describes as a deep commitment to the future. "This sort of profound feeling that I have, I don't want to romanticize it, but there's something, it is kind of an act of faith to bring a child into the world with no expectations. Even if this world is ending, even if we're marching through it, I can't surrender this idea that I'm not going to give up on the future. It's really important to me. As the saying goes, when the world is ending, what do you do? You plant a tree."

Monica's sense of commitment to the future is, ironically, based on having gone through war as a child. "You never know how what you do will work out. For me the historical trajectory has already been damaged. For me, the world has already ended when I was ten because I went through the war and that was pretty harrowing. It's like, the post-apocalypse." This postapocalyptic perspective—that the worst can happen and there is still life on the other side—is what drives Monica forward. Same with her commitment to leaving a legacy. "The idea that it's right now when we have to stop living, or stop reproducing, is sort of strange to me. . . . In some ways, this is the biggest risk I could ever take," she says. "I don't want to romanticize it, but there's nothing that makes you more human than that." She began to cry, quieting for a few moments. "But that's the vision of humanity. We have children because our time here is limited and there's something that continues on beyond us. It's kind of a way to push back against the tragedy of our existence. And so it's a deep desire to see something remain, whatever that is."

This is reproductive resilience in action—a commitment to creating, and leaving, a positive family legacy through parenting in this moment—and these conversations must be had out in the open, and often. Now is an important time for more people to situate their experiences with pregnancy, reproduction, and parenting in the context of climate change from a place of resilience, not shying away from the problem but facing it head on. This is not just about managing one's climate fears and concerns; it is about helping prepare children to be climate resilient as well. To do that, parents are going to have to have "the talk" with them.

THE NEW "TALK"

In many Black families in America, parents do not skim over or ignore difficult conversations, particularly about things like racial discrimination or the dangers of racial profiling or violence at the hands of police. Instead, they have "the talk." They prepare their children for what's to come. They let them know how to prepare themselves to interact with others and navigate a broader society shaped by racial inequality, even as they commit to surviving and thriving in difficult circumstances. It's going to become necessary for parents to have "the climate talk" with their children, and not just once, but over and over again as climate impacts intensify. This can look a number of ways, but it should begin with acknowledging the reality that climate change exists: that the Earth is changing and that those changes are leading to the unexpected ongoing challenges around us. These challenges include floods and fires, heatwaves and storms, and species disappearing. Teachers are having these conversations with children in school, but they tend to rely on scientific facts without an awareness of how those facts may land emotionally, which makes it all the more important for parents to also have the discussion at home.

These discussions are important because research shows that children are learning about climate change in school, they have feelings about it, and their more distressing climate emotions become worse if the adults in their life don't acknowledge the problem (or worse, if they dismiss or deny it).[26] A study looking at three hundred young people aged fourteen to twenty-four across the United States found that 75 percent of those in the survey said that the environment affects their mental health; one in four said that an environmental event or natural disaster had impacted their ability to concentrate in school.[27] Having the climate talk at home should begin with acknowledging the broad range of emotions they may be feeling, from fear and grief to motivation and determination. Asking what they know about climate change, how they feel about it, and acknowledging and addressing these emotions helps to make space for children and young people to talk about their experiences. Ensuring that they feel safe during these talks can help build their sense of safety and connection as they identify and navigate their feelings further through the climate crisis, now and in the future. And there are tools for doing this, such as the Climate Mental Health Network's guide for parents and caregivers, called "How to Talk to Young People about Climate Emotions."[28]

Tools are also available to help build emotional resilience to climate change in parents and nonparents alike. One example, developed by the See Change Institute, outlines the seven core components of climate-emotion resilience:

1. Acknowledge and validate feelings.
2. Draw on emotional coping tools.
3. Focus on social connection.
4. Connect with nature.
5. Take climate action.
6. Practice self-care.
7. Maintain a climate justice awareness.[29]

Resources like these cross the boundary between private and public action by engaging self and others at the individual, group, community, and society-wide level. It will take all of these actions and more—from coping with our own anxieties to holding our lawmakers accountable—to actually turn the tide and to help parents, and parent hopefuls, feel confident raising children through the climate emergency.

Reproductive refusers, resisters, and those who are resilient can contribute to making a better, safer planet. Nonparents can be, and often are, important members of family and community networks; reproductive resisters can raise awareness of the conditions under which having babies becomes viable or not, safe or dangerous, pointing the way to push for social and political change. And reproductive resilience is available to those who do move forward with raising children—it has always been required of marginalized people, and will become more and more necessary for those who are becoming increasingly vulnerable in new ways.

2 Can We Still Have Babies?

Carla's home is located on a dirt road on a ranch, in a small, unincorporated town in Southern California. At the start of the Covid-19 pandemic, she became a mother to a surprise baby boy. She was twenty-nine. Carla's pregnancy was difficult; she was often sick, which heightened her pandemic anxiety and raised concerns about the health of her baby. Reproductive services were curtailed and required a high level of vigilance; furthermore, getting to the doctor was difficult as it required a long journey over a bumpy road. Her little family lived in a trailer where the breeze whistles through the door hinges and around the window jambs. They had infestations of roaches and other pests. Winter nights got down to the thirties, mild compared with the more extreme temperatures of the Midwest or East Coast but uncomfortable nevertheless, as there is no central heat in the trailer. Summers often reach triple-digit temperatures, bringing a different kind of discomfort.

Carla's pregnancy was dehumanizing enough for the usual gendered reasons—that is, doctors second-guessed her at every step,

telling her she was eating too much when in fact she was barely eating at all due to sickness, and ignoring her requests throughout the
labor and delivery process. She's clear that this is because she is a
low-income woman of color: being Mexican American, she's confronted stereotypes about women like her having too many children.
Being pregnant was also frightening, as it took place entirely within
the broader context of a major public health crisis, which also mani-
fested as a crisis of social, political, and health systems. However,
these weren't Carla's main concerns during her pregnancy, nor during her early days as a new mother. Her primary concerns were for
her son and his future—and they were environmental.

"You know, it's just fear of the future environment, and what could
happen," Carla said. "What if all of the fish die out? What if the bees
die out? If there are no trees left on the planet, if the wealthy decide
to leave and move to the moon and leave the rest of us here with
what they've created, you know . . . so it's just . . . I don't think anyone should have kids unless they really, really, *really* think about it
and have a lot of support and they're a hundred percent sure that's
what they want. But still, I don't recommend it for anybody." Carla
tells me that she warns her friends against having children because,
as she learns more about environmental impacts, she worries about
whether she'll be able to keep her son safe as he grows up: "After
having my baby, thinking about smog and smoke and air pollution
like what I grew up with, and how it is going to be worse, it was like,
whoa! I don't know enough about this, and I wouldn't know how to
protect my child."

Carla's experiences and concerns encapsulate the core themes of
the chapter: the uneven ways climate change and broader crises
reshape the meanings and experiences of reproduction, pregnancy,
and parenting. Although individuals are tasked with managing these
issues on their own, they are not private problems. They are social
problems, revealed in the absence of appropriate state provision of
health resources, high-quality care, and other services to help families thrive. They are matters of reproductive justice.

Reproductive justice was developed in 1994 as a social movement and framework for understanding and responding to matters of reproductive oppression—specifically, the systematic ways low-income women, women of color, immigrants, queer people, and other marginalized groups experienced the incursions of the state in ways that curtailed their reproductive autonomy. Throughout US history, matters of reproductive capacity, or pregnancy, birth, and all that come along with them, have been matters of racial and class inequality.[1] Women of color and the poor have systematically suffered coerced sterilization, abusive medical experimentation, and restrictions to contraceptive and abortion access, at the hand of the state or through state-sanctioned policies. Resistance to these abuses has long been a central concern of the movement. At the same time, reproductive justice advocates identify justice for birthing people, parents, families and communities as being able to raise children in safe and sustainable environments—both in terms of being able to keep your children in your home and community, and to have that community be a place where those children will not only survive but thrive.[2]

Reproductive justice is holistic and comprehensive, moving from the individual to the family, the community, and the society at large, because it is necessary to understand that characterizing issues of pregnancy, birth, and parenting as matters of individual, private decision-making obscures how state power shapes these capacities at all level through laws, policies, or lack thereof. This chapter exposes the role of the state and its failures, arguing that dealing with reproductive anxiety and the kid question demands a reckoning with larger structural systems and failures. It demands a refusal to accept state failures as personal failures. And it demands that more people see themselves as participants and allies in the fight for reproductive justice, because the injustices of reproductive oppression are intensifying—and will intensify further in the climate crisis.

This chapter analyzes these issues at multiple scales. It looks at the international scale, where often inaccurate ideas about population growth and its role in greenhouse gas emissions drive policy

making focused on reducing human numbers rather than supporting the ability to create wanted families; nationally, in which the lack
of government-provided services makes it more difficult for people
to parent, even as the stakes of parenting are higher and harder to
manage; and in communities, where climate change is increasingly
impacting pregnancy and birth outcomes in uneven and unjust
ways. The early days of the Covid-19 pandemic offer instructive les-
sons as well, as they highlight how the multiple failures of public
service provision pushed the management of the crisis onto individuals and families, intensifying the burden of women's labor in the
household. The pandemic has revealed how, in a crisis, multiple disruptions at all levels land particularly hard on reproductive questions: In the face of a major crisis, can we still have children?

THE POPULATION BUST

In May 2021 the *New York Times* published a series of articles about
population decline in the United States. Citing demographers and
economists, they found that the birth rate had declined for the sixth
year in a row, bringing it to a nineteen-point decline since 2007.
There was also an overall decline in the birthrate among women in
their twenties, down 40 percent compared with 2007. The decline in
teen births was even steeper, a 63 percent drop compared with 2007.
Although some might point to a temporary baby bust in 2020,
reflecting a drop in new births amid the Covid-19 pandemic, that
trend was part of a longer downturn: it was the sixth year in a row
that the birth rate had declined.[3]

This reflects a much longer trend. According to the US Census,
the country's rate of population growth is the second slowest rate
since 1790, when the government first began counting births, both
because of the continuously declining birthrate as well as a decline
in immigration and an increase in deaths.[4] The high death toll is an
aberration; much of it can be attributed to the effects of Covid-19,

which claimed approximately one million lives from early 2020 to early 2022. However, the startling decrease in immigration is also significant. Between 2016 and 2021 total immigration to the United States plummeted by 75 percent.[5] Immigration hotspots like California are losing people at faster rates than other states, in addition to the decrease in new births. In fact, live births in Los Angeles have been decreasing so quickly that demographers project that as of 2020, births in the city will have decreased by 50 percent this century.[6] This decrease has clear political implications: California lost a congressional seat for the first time in history in 2021 after the 2020 census revealed the extent of the state's decline in population growth.[7]

All of this points to the fact that the American population bomb—the idea of explosive, unrelenting, and dangerous population growth predicted in the 1960s and promoted by environmentalists ever since—is a bust. This is true on a global scale as well. A 2020 paper funded by the Bill and Melinda Gates Foundation found that the global total fertility rate (TFR) will be 1.66 by the year 2100. This number, a measure of the total number of children a woman will have over the course of her lifetime, points definitively toward the fact that on a global scale people will not be replacing themselves. Population will be in a global state of decline, peaking in 2064 at 9.73 billion people and declining to 8.79 billion by 2100.[8]

What does all of this have to do with climate change? Actually, quite a bit: greenhouse gas emissions, the primary driver of climate change, have been steadily rising around the world at exactly the same time that population growth rates have been decreasing.[9] This counters prevailing dogma that more people equals more emissions. This point was driven home during the earliest months of the Covid-19 pandemic, when people around the world were forced indoors by home-stay orders, and global greenhouse gas emissions plummeted.[10] The plummet was not because there were suddenly fewer people on the planet overnight; rather, there were fewer people traveling by road or air. In other words, it was what people *did*, not *how many of them there were*, that caused carbon emissions to drop. And it is what people continue to do now—unsustainably consume resources, use polluting technologies, and engage in military warfare—that cause emissions to grow despite the long-term population decline in the rate of population growth.

Nevertheless, while mainstream environmentalists celebrate population decline, demographers, economists, and political leaders see the trend as deeply worrisome. In the way that current economies operate, nations that shrink rapidly cannot sustain their economies; elderly members of society don't have enough younger workers to support them in their old age. Anxiety about national geopolitical power grows as nations consider the balance of power shifting in relation to friends and foes. This sentiment also blurs into the fringes of environmental extremism, where racist ecofascists reside. A common fear among ecofascists is that the birth rate of the white population is plummeting in comparison to those of racial minorities, a "great replacement" that will cause social, cultural, racial, and environmental extinction.[11] As a result, ecofascists justify racist violence in the name of environmental and cultural preservation. As I demonstrate in chapter 6, ecofascists are dangerous not only because of their hateful rhetoric but also because of their ability to hijack the narrative on climate change and reproduction by turning the focus to population. In doing so, they reinforce long-held arguments that blame people of color and the poor for climate change, unsustainable resource consumption, and other systemic challenges.

In response to decreasing birth rates, countries like China, Japan, Russia, South Korea, and Turkey have shifted their national policy contexts to support birth. In China, where the country is projected to soon enter a period of negative growth, the infamous One Child Policy was reversed in 2015 in an effort to stimulate more births. In a country where total births per year are falling by the millions (from 17.9 million births in 2016 to 12 million in 2020), and there is a gender gap of 35 million more men than women in the country, policymakers fear a looming demographic crash in which there will soon be too few workers to support the large, aging population.[12] Some cities have gone as far as creating state-based online dating and matchmaker events for young Chinese singles.[13]

Other countries have taken more assertive measures: Russia offers monthly financial allowances for each child born, and bonuses on birth or adoptions of second children. Turkey offers incentives including free fertility treatments, financial payments after childbirth, and maternity leaves up to six months. France offers paid parental leave for up to three years, along with child allowances. Sweden offers a monthly check for each child up to the age of sixteen.[14] In South Korea, where the birth rate is 0.92, the lowest rate in the developed world, the government offers child allowances and fertility subsidies, and sends food, clothes, and toys to parents of newborns.[15] Even private companies are in on the effort. In Denmark the travel agency Spies created a series of ad campaign videos in 2014 and 2015 encouraging young Danish couples to use their travel agency to take romantic trips abroad—and come home pregnant, thereby helping to boost the national birthrate.[16]

The United States is *not* on the same track. The government isn't providing free childcare, new financial incentives, or guaranteeing paid long-term maternity leave to parents—in other words, they are not making it any easier for people to have children. At the same time, the Supreme Court has rolled back reproductive autonomy at the national level by overturning *Roe v. Wade* and handing abortion access rights to state leaders. This move will lead to more babies being born due to forced birth. At the time that I am writing this (in September 2023), most abortions are banned in fifteen states and restricted in nine others following the *Dobbs* decision; in many other states the struggle over abortion rights is currently taking place in courtrooms. In some cases, advocates are suing to block enforcement of strict restrictions, while in others, they are working to secure expanded legal protections for pregnancy termination.[17] In other words, in the face of virtually no federal support for parents, particularly working parents, the rollback of Constitutionally protected abortion access will actually boost the number of unsupported parents.

How can we make sense of this? Scholars, activists, and pundits have puzzled through this question for years, citing religion, patriarchy, and an American cultural emphasis on individualism and self-reliance as reasons for restricted reproductive rights on one hand and a general lack of financial support of parents on the other. It also comes down to politics, economics, and how they come together. Jenny Brown's analysis of reproductive labor, *Birth Strike: The Hidden Fight over Women's Work*, theorizes that efforts to roll back abortion and contraceptive access in the United States are actually in line with broader government goals to boost the number of people who will contribute to the economy by working, paying taxes, and using their dollars to consume—and that families are forced to pay for the creation of this boosted workforce themselves.[18] As she notes, a higher birth rate caused by the rollback of reproductive autonomy "serve[s] an economic goal: an ever-expanding workforce raised with a minimum of public spending and a maximum of women's unpaid work."[19] The result of all of this is more stress on working individuals, and less time, money, and availability for childcare and other caretaking roles.[20]

This is facilitated by what sociologist Caitlin Collins refers to as the "principle of personal responsibility" that undergirds so much of mainstream American culture.[21] With no explicit national family policy, no universal programs for work and family, and no requirements for employers to provide them, US society organizes family life based on a free market–based approach that says that we must work and find private strategies for addressing childcare and housework. In this logic, having children is a "choice," and those who make that choice should only do so if they can afford it.[22] In the face of new restrictions to abortion access, low-income women—a disproportionate number of whom are women of color—are more likely to experience unwanted births because of an inability to travel to other states to maneuver around restrictive abortion laws. This is a problem, because being a woman of color in the United States increases the likelihood of experiencing a range of dangerous pregnancy and birth outcomes, from pre-term birth to low birth weight. So does climate change.

ENVIRONMENTS, EXPOSURES, AND REPRODUCTIVE INJUSTICE

Pregnancy, birth, and the postpartum period can be risky under the usual circumstances. Add hotter temperatures, worsening air pollution, and stressful eco-anxiety to the mix, along with forcing people to be pregnant and give birth, and the combination can be downright dangerous. As the planet warms, air pollution and other toxic environmental exposures are ever more concentrated, particularly in urban centers, and it becomes harder and harder to make and sustain a pregnancy. Without addressing how environmental changes impact human reproduction, the question of whether we *want* to have children or not will be overshadowed by whether or not we *can* have them.

Environmental factors pose direct risks for pregnant women and babies. For example, birth outcomes like stillbirth and preterm birth are associated with air pollution.[23] Preterm birth is also significantly correlated with mothers being exposed to higher temperatures during pregnancy. Babies born too early have higher risks of death and disability; the final weeks of pregnancy are important to the development of key organs, including the brain, lungs, and liver.[24] Preterm births (births before thirty-seven weeks of pregnancy) are highest among Black women, whose rates are 50 percent higher than the rates for either white or Hispanic women—and preterm birth is significantly more common among mothers who are exposed to higher temperatures during pregnancy.[25] One study conducted in 2019 found that, on average, pregnant mothers had lost 151,000 days of gestation each year in the United States due to heat exposure. The authors estimated that "at a minimum, just under 25,000 births per year occurred earlier than otherwise," and that this gestational loss is greater for Black mothers.[26] Although Black women disproportionately suffer these challenges, the effects of environmental change will impact all pregnant people, particularly as heat waves and increased temperatures become more common across the nation.

Another dangerous outcome, low birth weight—which is associated with higher rates of infant mortality, childhood respiratory problems, and psychological challenges in adulthood—is linked to mothers' exposure to air pollution, in addition to behavioral exposures and chronic health issues. Rising temperatures also increase this risk. A California study of more than two million babies born between 1999 and 2013 found that long-term exposure to higher temperatures, particularly in the second and third trimester, was a factor in low birth weight. This association was highest among Black women, older women, and those who gave birth during the warmest months.[27]

Despite what scientists know about the increased danger of heat events for pregnant women, leaders are slow to share this information with the public. A *BuzzFeed* investigation reviewed the heat and safety guidelines of emergency and health departments in twenty-five of the most populated cities in the United States and found that only two of them—Chicago and Philadelphia—included pregnant people in their heat warning. This was made even more stark in the early days of the pandemic, before the availability of vaccines, when the twin challenges of intense heat and the threat of exposure to the virus combined to heighten the vulnerability of pregnant women.[28] More disturbing is the fact that some officials are hiding or obscuring their heat death data to make the toll seem lower than it is. For example, a 2021 exposé in the *LA Times* revealed that the state of California significantly undercounted its heat-related death toll from 2010 to 2019, attributing 599 deaths to heat exposure when the toll was actually more than six times that number.[29] Given that extreme heat is one of the deadliest impacts of climate change, accurate tracking and reporting of the consequences of heat events is a life-and-death issue. Given the racial health disparities around pre-term birth and low birth weight, it is also an issue of reproductive justice.

Lawmakers are increasingly becoming aware and proactive on the linked issues of racial justice, reproduction, and climate change, and some have responded by introducing a range of bills to address maternal health disparities.[30] In one example, a congressional proposal acknowledged how climate change–related heat, air pollution, weather-based disasters, and other risks negatively impact pregnant and postpartum women and their babies. The proposal includes allocating research funding to further identify how climate change is associated with high-risk maternal and infant health outcomes.[31] These moves are an important start in the right direction; a concerted effort to understand and intervene on the relationships between extreme heat, air pollution, and pregnancy and birth outcomes in general is required to understand the conditions that will affect more and more people with each passing year.

Environmental impacts on pregnancy and birth outcomes are expanding as heat events and other climate impacts continue to intensify—and in return, more and more women and their children are becoming increasingly vulnerable. For women of color, these impacts are clear and on the surface: racial health disparities ensure that health burdens show up first and worst among communities marginalized by racism and poverty. However, there is also an opportunity here for more people to align with women and people of color across race and class divides to seek reproductive justice. As climate change intensifies, it will make more pregnancies vulnerable. It will threaten the ability to raise wanted children in safe and sustainable communities, as called for in the reproductive justice framework. The Covid-19 pandemic demonstrates how a crisis can expose state failures and broader vulnerabilities for reproducing people. It demonstrates why we need a much broader coalition of participation and support for all aspects of reproductive justice.

THE COVID-19 CRISIS

Although Covid-19 is different from climate change, the pandemic is instructive in thinking about the kid question, because it has had a significant impact on sexual and reproductive health, childbearing, and parenting in the United States. When Covid first hit hard in the late winter and early spring of 2020, women and couples who wanted to have babies found themselves facing a significant amount of economic and emotional strain, as people lost jobs, were saddled with new medical debt, and were afraid to leave home. When they did brave the outside world, they faced new barriers to accessing medical care.

For example, a national study of 6,211 women conducted in 2021 found three key things: first, the pandemic created a barrier to accessing sexual and reproductive health care, including contraceptive access, which led some people to change their intentions around having children; second, the pandemic had ongoing, disproportionate impacts on women's sexual and reproductive health, particularly for members of socially marginalized groups; and third, telehealth services stepped in to bridge the gaps, particularly for those who already experienced other barriers to accessing care. Of this group, 15 percent wanted fewer children or to have children later because of the pandemic (the top reasons were financial concerns; state of the world; and fear of being pregnant or giving birth during the pandemic); one in five had to cancel or delay a doctor's visit for sexual or reproductive health care, or had difficulty accessing their contraceptive method during the pandemic. These barriers were strongest for women of color, LGBTQ women, and those of lower income.[32]

It was a scary time. Women in places with high case rates had to give birth alone when their partners, doulas, and other support companions were barred from accessing delivery rooms; fertility clinics suspended treatments, in some cases indefinitely; and there were concerns and some evidence that Covid-19 could be transmitted in utero.[33] Of course, not all of the concerns were biological: the pandemic was devastating for mothers of school-aged children. Women lost approximately sixty-four million jobs worldwide in the early days of the pandemic; of those who didn't lose jobs, the majority saw their workload increase as their optimism about their career prospects declined.[34] Household chores, which increased dramatically in a short period of time, fell disproportionately to women, as did childcare and supervision of children's virtual schooling. This extra burden of gendered labor fell on women's shoulders, regardless of income; the disparity was worse in households where men worked outside of the home during the pandemic.[35]

In *Academic Outsider: Stories of Exclusion and Hope*, sociologist Victoria Reyes captures the everyday impacts Covid-19 has had on the lives of professional women.[36] "The Covid-19 pandemic has collapsed any sense of boundary between work and home life," she wrote. "Instead, it has demanded that mothers and other types of caregivers work *overlapping shifts*. Each and every responsibility demands attention at the same time. The roles overlap, bleeding together and leaving caregivers—myself included—unable to really do any job well. I feel like I'm failing at everything, whether it's the research, teaching, or service demanded by my paid employment, or the elder care and childcare needed at home." The pandemic revealed that despite decades of women's advancement in education and employment, household labor is still unevenly divided by gender, creating burdens that continue to fall primarily on the shoulders of women and mothers. It revealed, in a larger sense, that when the state fails to provide necessary social services, including reliable, high-quality health and medical services, individuals, families, and communities are left to fill in the gaps and bear the brunt of systemic harm.

And yet many people do still have children, even while navigating pandemics, climate change, racism, social and political polarization, and all of the rest of it. For them, reproductive resilience is a requirement, but it is not enough. Only collective resistance to the systemic harms of a state that rejects reproductive autonomy can work to connect the micro-level forms of resilience needed to address the climate anxiety-driven kid question with the macro-level systemic issues driving climate crisis and reproductive health crises. We need more people across social categories of difference to see themselves in the fight to raise their children in safe and sustainable communities. We need more people to join that fight, recognizing that a struggle against reproductive oppression benefits everyone. In other words, we need reproductive justice for all.

3 The Kids Are Not Alright

July 2019. It's a lazy summer day in West Los Angeles traffic. My mother and I are on our way to Culver City for our favorite leisurely Saturday mother-daughter date: a massage, followed by a visit to the restaurant next door for hot, steaming bowls of pho and cold glasses of Vietnamese iced coffee. The conversation meanders, eventually coming to my research and this book. My mother asks what the central message of the book is, but the project is in the embryonic stages, so I don't have a ready answer for her. I think for a while, and finally turn to her. "It's about why you might not have great-grandkids in the future." My nieces are six and eight years old; my mother sits upright and stares, openmouthed.

"Well, I've told you that the book's about reproductive anxiety and how young people feel about climate change," I tell her. "Yeah . . . ," she says slowly, not quite sure how to connect the dots. "I actually feel very anxious about climate change," I say. "I'm not sure this Earth is going to be a very comfortable place to live twenty or thirty years from now. I mean, it's not the reason I don't have kids, but it's

definitely part of the picture for a lot of people." She looks out the window, silent, thinking about my nieces. I explain to her some of the results of recent climate emotions surveys, how children and teenagers are worried about the planet's future. "You'd be surprised," I tell her, "but this is a really big issue for a lot of young people. They're questioning whether we have a future ahead of us at all."

My mother remains silent for a few moments, so I fiddle with the radio station and check the rearview mirror, assuming she's moved onto another thought. Finally, she whispers: "It's so unfair. Why should they have to think about that?" I glance her way and feel a twinge of guilt. Maybe I shouldn't have upset her. Just as I start to craft a convoluted apology/retraction/feel-good salvo, she brightens. "Wait a minute! You're talking about white people, aren't you? I know Black people aren't talking about that. We don't think that way. Was this research done on Black people?"

She waits for a response, but I have nothing to say. I've only completed a small handful of my own interviews at this point, and I don't have the numbers that would convince her that people of color experience climate and reproductive anxiety. My mother has a point. The few surveys I've seen have focused on white people to the near total exclusion of people of color. It's left me wondering the same thing: is climate and reproductive anxiety a white thing? It couldn't be, because people I knew were expressing these concerns. I had these concerns too, which I had never actually discussed with anyone before this project. If I wasn't talking about it, that meant others weren't either. And some of those others had to be people of color too.

As evidence of climate crisis has grown, opinion pollers and academic researchers have met the moment. Since the turn of the millennium, there has been growing research on people's feelings about climate change. More recently, however, one particular group has received a lot of attention from survey researchers: Millennials and Generation Z. These surveys focus on their reproductive plans and behaviors, reflecting a broader interest in how climate change is shaping broader population trends, both now and in the future. But this is generally not a racially diverse group at all. Instead, people of color are either excluded from this research or recruited in such low numbers that it's impossible to draw any conclusions about the significance of race. In 2021, I conducted a national survey as a corrective to this race-blindness in the research. I share the results of that survey in this chapter, to demonstrate why we need to pay attention to race as we learn more about reproductive anxiety and the kid question.

THE REPRODUCTIVE ANXIETY GENERATION

Climate-driven reproductive anxieties are a widespread concern. These are generational issues impacting those in the thick of their reproductive years—mainly Millennials and Gen Z—but these anxieties are also salient because these generations grew up with knowledge of climate change and its impacts from earlier ages than prior generations. They also have a general awareness of the long-standing impacts climate change will have across their lifespans as well as those of any children they bring into the world.

In 2018 the *New York Times* surveyed men and women ages twenty to forty-five to find out why fertility rates are declining in the United States. The results were that of those who had, or expected to have, fewer children than they considered ideal, one-third cited climate change as a reason why.[1] Several years later, the blog Modern Fertility polled over twenty-eight hundred people; more than half said that they were considering either having fewer children or not having children at all because of climate change.[2] A quarter responded that they were considering adoption, and another third said that they had considered moving to a city or state with fewer climate impacts. This was particularly true for people in and around Los Angeles, San Francisco, and New Orleans—places repeatedly affected by wildfires and devastated by hurricanes.

Another poll was commissioned the same year by cleaning products company Seventh Generation. The survey of two thousand Americans was evenly split across generations: Generation Z, Millennials, Generation X, and Boomers. The survey found that 75 percent of Gen Zers and 77 percent of Millennials said that climate change has shaped some of their major life decisions.[3] Among Gen Zers, 78 percent reported that they aren't planning, or don't want, to have kids as a result. Across the board, 59 percent of those surveyed reported that climate change has negatively impacted their mental health. The survey did not look directly at the relationship between negative mental health and reproductive plans but did suggest a possible connection.

One challenge with these broader, nonscientific polls is that they don't tell us much about smaller groups within the population. This leaves us to assume that demographic characteristics other than age—like race, for example—don't matter, because reproductive anxieties are universal. However, when data about race are included, the vast majority of those polled are white, which would seem to suggest that the climate and reproduction concerns of white people are generalizable to the broader public. Neither is necessarily true. We don't know whether race matters in climate-and-reproductive anxiety or not, and we *can't* know unless the people who create the surveys and other studies include questions about race, and include enough racially diverse people in their samples, to see whether there are meaningful differences in how they respond.

This kind of information did come forward in a national survey of forty-four hundred people across ages eighteen to over sixty-five, conducted by Morning Consult in September 2020.[4] Those in the survey who did not have children were asked, "To what extent are the following a reason why you don't currently have children?," followed by a series of answers, including financial challenges, concerns about Covid-19, and other issues, such as climate change. Overall, the answer, "I am concerned about climate change" garnered little response; 75 percent of respondents said that climate change was not a reason for their decision at all. However, this changed in the different age categories: 15 percent of those eighteen to thirty-four said that climate change was a major reason for their decision, while 21 percent said that it was a minor reason. These numbers were similar for those thirty-five to forty-four years old.

Probing a little further, an interesting racial-generational picture emerged: while only 21 percent of white respondents said that climate change was a factor in why they did not have children, 41 percent of Hispanics and 30 percent of Black respondents said that it was. This suggests race is a factor in how climate change shapes the kid question, but how? What's race got to do with it?

RACE, CLIMATE EMOTIONS, AND REPRODUCTION IN THE UNITED STATES

In July 2021, I began developing my own survey. It was designed to include people across the United States and to look at the relationships between race, climate emotions, and people's desires and plans for having children, with a focus on people between the ages of twenty-two and thirty-five. The average age at first birth in the United States is twenty-seven years old.[5] Thus choosing this age range would ensure that these would not be solely hypothetical questions, but rather something people in the survey are actively thinking about, making plans for, and carrying out. I developed and launched the survey through a national survey organization and set the inclusion criteria in two ways: participants would have to have a high school diploma and answer yes to the question, "Is climate change real?" Although I knew that this would exclude some people, it was important to only capture those who accept that climate change is a reality, thus making clearer whether their emotions and reproductive plans and behaviors are a response to it. The target number of participants was twenty-five hundred, with half of the responses from people of color, and the rest white respondents.

Survey respondents were 2,521 individuals representing a broad, diverse range of people who are worried about climate change (84 percent).[6] Their concerns, in this order, were wildfires, heat waves, strange weather, hurricanes and other storms, and sea level rise. On the whole, their emotions in response to climate change were (in this order): concerned, uncertain, anxious, powerless, scared, sad, overwhelmed, and depressed. When asked about their emotions toward parenting (whether they have kids, plan to have them, or not) in the midst of climate change, their responses were mostly the same: uncertain, concerned, anxious, scared, and—perhaps unexpectedly—hopeful.

From there I looked at racial differences in the responses, and that's when it got interesting. Respondents of color were significantly more likely than white respondents to report optimism and hope with respect to climate change in general, and all of the positive emotions (optimistic, hopeful, motivated, determined, happy, and excited) with respect to parenting in the midst of climate change. With respect to climate change in general, respondents of color were significantly less likely than white respondents to report feeling angry, resentful, depressed, powerless, overwhelmed, and uncertain, or to report feeling checked out or having no feelings. On that last note, white men were the most likely of anyone in the survey to report numbness about climate change. People of color were significantly more likely to choose one intense negative emotion: traumatized.

With respect to *parenting* in the midst of climate change, respondents of color were significantly less likely than white respondents to report feeling angry or uncertain, or to report having no feelings at all. Religion had an effect on climate-parenting emotions across the board, with respondents identifying as Christian generally being more likely than other respondents to report feeling positive emotions (optimistic and hopeful with respect to climate change in general; optimistic, hopeful, motivated, and happy with respect to parenting in the midst of climate change) and less likely to report feeling negative emotions (sad, anxious, scared, uncertain, concerned, and confused with respect to parenting in the midst of climate change). Respondents of color, particularly women, were more likely to be religious Christians.

Ironically, the determined, motivated (some would say more positive, action-oriented) emotions did not translate into people of color's intended parenting plans. When asked how many kids they would want if climate change was not happening, most people in the survey overall indicated that the number would not change from whatever number it is now. However, for those who said they want—and otherwise would have—more children if climate change was not a factor, twice as many were people of color. In the data analysis, it became clear that this was not a function of income, education, political identity, age, religion, gender, or any particular emotion category; it was a function of race. People of color were considerably more likely (with statistical significance, or the likelihood that you are actually observing a real difference, and not a coincidence) than white people to take climate change into their future reproductive plans and to plan to have fewer children as a result.

There was only one thing that made it more likely for a person to respond this way: race. To learn more, a colleague and I ran an analysis to uncover two specific things: which emotion and which demographic characteristic most determine who plans to have at least one fewer child than the number they actually want, because of climate change? The responses were clear and statistically significant: the most important climate emotion was *fear;* the most important demographic characteristic was *race.* So it was clear: race is a key factor in Americans' climate-driven reproductive anxiety and is shaping some young people's plans to have fewer children. But why?

RACE MATTERS

The survey was quantitative, focusing on collecting numerical data based on limited, closed-ended questions. It did not ask the kind of open-ended questions that would draw conclusions as to why race matters in climate emotions and reproductive plans. However, based on the quantitative results and the broader literature, I want to offer some possible explanations of what is happening. These explanations are grounded in a climate justice perspective, one that acknowledges the unequal distribution of climate burdens to low-income communities, primarily communities of color. And they suggest that climate-driven reproductive anxiety may be an important, but largely hidden, way that climate injustice harms our communities.

Explanation 1: People of color feel generally vulnerable in the United States, and as climate change makes physical vulnerabilities worse, it extends into how they feel about being able to create the families they want.

As we saw in chapter 2, racism in the United States has deep historical roots that result in physical health impacts and racial health disparities. Beyond maternal and child health, racial health disparities include the unequal death rates that ravaged communities of color during the Covid-19 pandemic; rates of diabetes, cardiovascular disease, and respiratory problems like asthma are also higher in low-income communities and communities of color. Violence poses yet another life-threatening risk: Black people in the United States are more than twice as likely to die from gun violence compared to whites or Latinos, and three times more likely to die in an encounter with the police.[7] Native Americans are even more vulnerable; despite their relatively small numbers, they are statistically more likely than any other group to die at the hands of the police.[8] Circumstances are particularly dire for Native women: thousands of Indigenous women and girls go missing and are murdered every year, but most of these cases are never investigated by law enforcement. In 2016 alone, while 5,712 Native women were reported missing or murdered, the US Department of Justice only logged 116 of the cases in their database.[9] Shockingly, murder is the third leading cause of death for American Indian/Alaska Native women.[10]

This general sense of social and physical vulnerability is also linked to the environment. Environmental and climate justice research has shown for decades that systemic and pervasive racism—as well as histories of land theft and redlining—forces communities of color to into communities plagued by environmental problems, including exposures to toxins in air, soil, and water as well as urban heatwaves. For example, in Los Angeles, Black residents are nearly twice as likely as other city residents to die in a heat wave as a result of being segregated into heat island communities—places with lots of heat-trapping concrete and little tree cover, which concentrates heat—and because low-income community members have less access to home air conditioning, cars of their own, and cooling centers to escape the heat.[11]

This reflects a nationwide trend. According to an Environmental Protection Agency (EPA) study looking at census tracts and expected temperature change, Blacks are 40 percent to 59 percent more likely than other groups to live in the places most likely to have the highest death rates due to climate-induced heat events.[12] Five of the smoggiest cities in California have the highest concentrations of residents of color; these are places projected to have the largest climate change–induced smog increases, with negative impacts on respiratory and cardiovascular health.[13] Indigenous groups are 48 percent more likely than other groups to live in regions where more land will be inundated by sea level rise, and Hispanic or Latino groups are most likely to live where people will lose the most working hours due to extreme temperatures, directly impacting the financial health of families and communities.[14]

It is not a coincidence that people of color live in these places. Rather, this is a systemic pattern of environmental injustice. Research over decades has shown that siting and land-use patterns near communities of color expose these neighborhoods disproportionately to environmental harm. These impacts extend to mental health outcomes as well; a 2021 report from the American Psychological Association noted that the pervasiveness of American racism leads to negative mental health outcomes for communities of color; intensified climate burdens threaten to add even further to negative mental health outcomes.[15] General vulnerability due to systemic racism, compounded by climate vulnerability, likely factors into the kid question and leaves people unsure of their ability to confidently raise families that they can protect from harm in an intensifying climate crisis.

Explanation 2: People of color are more concerned about climate change, and this concern translates into a concern for their future potential children.

A 2019 Yale survey found that Hispanics/Latinos and African Americans care more about climate change than their white counterparts, and are more likely to be alarmed or concerned about global warming, while white Americans are more likely to be doubtful or dismissive.[16] A follow-up study conducted three years later found that 10 percent of Hispanic/Latino adults surveyed reported that they experienced clinically diagnosable anxiety or depression related to climate change; 21 percent indicated an interest in discussing their feelings about global warming with a therapist, compared with 10 percent of Black respondents and 5 percent of white respondents.[17] These responses may stem from a general feeling of vulnerability as described above or from direct observation and concern with climate impacts they've seen in their own lives.

The response may also stem from a concern about the financial costs of adapting to climate impacts, whether they are impacts on the air conditioning bill or the price of moving to a community less likely to be washed away in a flood. A research report found that low-income communities and communities of color may spend as much as a quarter of their household income on basic necessities like food, electricity, and water—all of which may become more expensive due to climate change.[18] At any rate, if communities of color are more concerned about climate change, it stands to reason that the same communities are also more concerned about the future conditions these changes will bring. Given that children are cherished symbols of the future, factoring climate concerns into reproductive plans makes sense.

Explanation 3: This isn't actually about people of color; instead, white respondents are less reproductively anxious than they should be. This is because social privilege acts as a buffer against feeling vulnerable to existential threats.

Britt Wray's 2022 book *Generation Dread* offers a thoughtful first-person account of how the author, as a young, middle-class white woman, became aware of—and quickly overwhelmed by—her emotions in response to climate change. The main reason? She had never experienced or contemplated an existential crisis before. One's social privilege, based on race, class, and other markers of vulnerability and marginalization, plays a significant role in how a person responds to learning about current and future climate impacts. Communities that have spent generations confronting systematic exposure to ill health, early death, loss of land and cultural sovereignty, and all manner of suffering along the way are familiar with existential threats—and for better or worse, these communities have developed coping mechanisms that may factor into the decisions they make about their families. Wray notes that she first confronted her climate fears through her struggle with the kid question—a process involving grief, sadness, loss, and depression. Although more and more white Americans are beginning to confront these questions, many may still have enough faith in social and economic institutions to believe that they will be protected from the worst harms to come.

THE UNBEARABLE WHITENESS OF REPRODUCTIVE ANXIETY RESEARCH

As this chapter shows, there is a small but growing body of scholarly research looking at climate-driven reproductive anxiety. Those studies consist primarily of a handful of surveys conducted by university research teams in the United States, Canada, and Europe looking at the relationships between environmental concerns and reproductive plans. Moving beyond the in-the-moment snapshot provided by general opinion polls, these studies offer more context for understanding the broader impacts of environmental concerns on reproductive anxieties. They don't, however, deal with race. In fact, the studies conducted in racially diverse societies are *far* more homogenous, and focused on white participants, than is representative of those societies. Not only do the studies not appropriately represent nonwhite people of color, some do not include racially diverse respondents at all. Some of the studies explicitly center neo-Malthusian ideas, blaming reproduction and population growth for climate change—a way of thinking that contributes to blaming communities of color and the poor for environmental problems.[19] In other words, either people of color are minimized, erased from the studies, or vaguely framed as environmental villains.

Take, for example, a 2012 study that looked at environmental concerns and reproductive plans among college students in Canada.[20] Researchers surveyed just 139 students ranging in age from seventeen to forty-four years old, 83 percent of them white. They wanted to test the hypothesis that people's ideal number of children is related to environmental concerns, specifically concerns about negative environmental impacts on human health, and also the negative ways humans are impacting the planet. That last part was measured based on people's responses to a tool called the New Ecological Paradigm (NEP), a scale that assesses environmental attitudes based on people's responses to statements like, "We are approaching the limit of the number of people the Earth can support," "Humans are seriously abusing the environment," and "The Earth is like a spaceship with very limited room and resources."[21]

The clear neo-Malthusian undertones in their measurement scale reveal the study authors' assumptions about population as a negative environmental force. However, the responses were not what they expected. The people who took the survey said that their health concerns about pollution's impacts on their physical and mental health were their biggest concerns in terms of wanting to have fewer children. The NEP tool, which measured how they felt humans were impacting the environment, had no direct relationship to people's reproductive plans. This suggests that people care a lot about how the environment will impact or even harm their health in the future—more so than they care about blaming human numbers for causing environmental harms—and that this concern is a key component of reproductive anxiety. But you can't disregard race, because how people respond to their concerns is shaped by how vulnerable they are to social inequalities.

Two researchers in Europe decided to build on the Canadian study, asking similar questions of a much larger group of people. In 2013 they studied how "worsening environmental conditions" impact fertility intentions among 8,278 people from twenty to forty-five years old, in every EU country in Europe.[22] The authors specifically wanted to measure how people who are sensitive to environmental changes, and those who are fearful that the future environment will be an unhealthy one, feel about having children. Their hypothesis was that these people's attitudes would favor either having fewer kids or having them later—but they were wrong. To the authors' surprise, those who were most concerned about climate change had more positive attitudes toward having children, and having more children, than others in the study. This held true across levels of education and country of origin, and it suggests that people may become more concerned about environmental issues like climate change when they already have kids or are actively planning for them. In other words, for some, children might move climate change from an abstract concept to a lived reality—one that they will need to become more aware of, and actively work to fight against, to protect their children. Unfortunately, no data on race was included in the study.

You might say, *Well, race and racial inequality are experienced differently across societies; racial vulnerability in the United States isn't something we can generalize to other places, and perhaps neither are reproductive anxieties.* True—but in that case, why don't United States–based researchers include race as a component of their reproductive anxiety research? In a society where social privilege and inequality are strikingly distributed along racial and ethnic lines, this should enter into the research, but instead, researchers have continued to exclude race from their analyses by recruiting tiny, nonrepresentative numbers of people of color—thus making any kind of racial comparison impossible. For example, a 2020 survey of 607 Americans found that just under 60 percent of participants were either "very" or "extremely" concerned about the carbon footprint of having children; 96 percent of them said that they were "very" or "extremely" concerned about the well-being of existing or hypothetical kids in the face of climate change too.[23]

To their credit, a number of respondents rejected overpopulation narratives and population control as a solution to climate problems. But 88 percent of the survey participants were white, 91 percent were politically liberal, and 93 percent had a college degree or higher. Almost all of them were women, and they were recruited to participate in the study based on their involvement with some of the largest environmental organizations in the country, which have a history of predominantly white membership. All of this tells us that, while interesting, studies of this kind of narrow demographic group—well-educated, white, politically left—tell us very little about what's actually going on in the broader society. And if we take studies like this as an indicator of who suffers from reproductive anxiety, we have a very limited view. We can't assume that the experiences of this very small group are generalizable to anyone else, which is why we need research on communities of color.

One last study, conducted comparatively in the United States and New Zealand, looked at the comments sections of articles about climate change and attitudes toward having children.[24] The researchers found that the people most likely to use the comments sections to debate the causes of climate change often identified population growth as the leading cause; they also used the space to justify their own childbearing choices, mainly through a racist lens worrying about "overpopulation" in the Global South and how it might threaten the security of Western nations. The authors did not analyze race in this study, of course, because people don't necessarily reveal their racial backgrounds in online comments sections; however, in a second part of the study, the researchers interviewed twenty-four young adults (twenty-one of whom were white) on the same topics they had analyzed online. Of the tiny group of participants, their concerns focused on the negative impacts of future children on the planet, specifically through overpopulation and overconsumption. Uncertainty and guilt were common feelings interviewees expressed. And all of them said that "not having children was the biggest positive choice one can make for the environment."[25]

Perhaps this last study does say something about race; the negative attitudes white environmentalists have toward *other* people having children, particularly toward those in the Global South. These are long-standing racial attitudes about population size and growth that blame the poor for environmental problems. These attitudes have a long, ugly history among white American environmentalists and have publicly resurfaced among ecofascists (discussed in chapter 6). Ironically, there have been horrific displays of ecofascist terror in the United States and New Zealand, the nations where the study participants live. And these kinds of racial attitudes, destructive as they are, are worth knowing more about so that we can better understand how environmental messaging might perpetuate them.

This chapter is a call for racial inclusion in research, specifically research on climate emotions and reproductive anxiety. In a society like the United States, where environmental exposures and life chances are heavily shaped by racial inequality, we just can't afford to conduct so-called "race-blind" research. We also can't ignore the intersections of race with other marginalized identities, particularly LGBTQ+ identity. A small but growing body of research shows that LGBTQ+ people face unique challenges during disasters, including exclusion from shelters and discrimination based on gender identity and presentation as well as nonrecognition of their families in emergency evacuation sites. These effects are strongest for LGBTQ+ people and families of color.[26]

At the time of this writing, the published research on climate and reproductive anxiety does not focus on intersections of identity including gender identity and sexuality, but it should. Scholarly research can replicate injustice through exclusion of marginalized individuals and communities, or it can do the opposite by providing an evidence base to prove that a problem exists. Excluding people of color from reproductive anxiety research creates a missed opportunity to understand a growing set of concerns among young people— concerns that will likely expand as the effects of climate change worsen.

4 The Kid Question Is Unjust

The kid question. It comes up over and over again in the form of family questions and expectations. It arises in conversations with peers, partners, and new dates. It appears in the quiet times, sitting in the spaces where our wildest hopes and deepest fears collide. This question is not small or lighthearted; it is not a thought experiment. It adds pressure to the task of figuring out how to create a future life that will be happy and safe, for self and loved others. Including others who don't yet exist.

These others are, of course, children. For me, questions from family members have gotten quieter and less frequent as I entered my forties and my older sister had children, alleviating some of the pressure that was previously directed my way. The internal questions have also quieted at the end of my reproductive window, but that doesn't mean that there aren't cherished children in my life. I am an auntie, a godmother, a fun older cousin. They're not a family of my own creation, but I do wonder about their future, how they will grow up, whether they will ever have families of their own, what the planet will look and feel like when the time comes for them to ask those questions for themselves. For you, those children may be actual babies gestated in the womb. Or they may be children born of others' wombs, families, and communities, brought in to complete the vision of a family you've otherwise had difficulty creating. Regardless of how they arrived, they are a family of your own creation. You are responsible for them. You must keep them safe, healthy, out of harm's way. It is your job to ensure that they survive to adulthood with all of the tools and resources necessary to help them thrive and be happy.

As you contemplate that, you wonder whether you are capable. Can you do this? Perhaps you've struggled with anxiety and depression. Should you have kids? American society feels more socially and politically polarized than ever—is it right to bring another person into that? Racism has shaped some of your key life experiences, which you would never want a child to endure—can you let another generation go through that? You worry about the next strong rainstorm or snowstorm after watching news footage of recent floods that killed dozens of people, or blizzards whose lingering blankets of ice and snow took more than a month to clear away. You think ahead to September and October, hurricane season, and say a little prayer that this year's storms won't be as bad as last year, when you had to evacuate, returning to find that your house had been hit even harder than you'd imagined. Or you look out of your window at a hazy August sky, smoke in the air and the temperature at a constant 102 degrees, your asthma getting worse by the year. You feel a sense of sadness seep in, wondering whether this is the new normal, or— even worse—if this is the mildest version of what's yet to come. Through it all, you think, *Can I have a kid in the midst of this? Should I?*

But then you spend an afternoon with your family, watching movies, playing with the dogs, or having a very earnest conversation with your three-year-old nephew about ducks, and you breathe a sigh of relief. Your family is imperfect but they're yours. And when you're with them, you get to just *be*. There's safety in it. Protection from the outside. Maybe even happiness. Or even the realization that their imperfections are exactly why having this for yourself is so important. So that you can break the cycles, create the safe space, be the person you needed when you were little. You think, *This is exactly what I want. I want a family of my own.*

These competing feelings—fear and anxiety about climate change, an insistent desire to create a family, and in some cases, the determination to break with family tradition and forge a different path—are the subject of this chapter. The chapter focuses on a series of interviews I conducted with Millennials and members of Generation Z in 2021 and 2022.[1] They are a diverse bunch, all of them people of color: Black, Asian, Latino, Native American. Some grew up in low-income families and neighborhoods while others' parents were middle or upper-middle class. Some of them identify as queer, or their close family members and friends do, which shapes their sensitivity to discrimination against gay, lesbian, bisexual, and transgender people. An awareness of, and sensitivity to, the intersections of inequality has played a big role in their lives, their families, and their visions of the future. In some cases, so have the experiences of immigration and adjusting to cultural and generational differences.

Their experiences as members of marginalized groups shape their experiences with climate emotions like anxiety, fear, and trauma—as well as hope and optimism. In this chapter, I argue that their challenging climate emotions are the result of injustice. These emotions fall hardest on those already plagued by a general sense of vulnerability, whether it's because they grew up in low-income families, attended under-resourced schools, or lived in communities that were heavily policed and racially profiled. It could be because they watched their immigrant parents navigate financial and social barriers, or because they know that the polluted air hanging over their neighborhoods doesn't plague the neighborhoods of people with more social privilege and wealth.

While everyone is subject to the experience of uncertainty, some people are protected by their privilege. A report analyzing the period from 1999 to 2013 found that white Americans and those with more wealth receive more federal money after disasters than people of color and low-income people. The report analyzed more than forty thousand records from the Federal Emergency Management Agency (FEMA) and found that 85 percent of post-disaster buyouts had gone to white, non-Hispanic families.[2] Other studies have corroborated this finding, demonstrating a clear pattern: in the wake of disasters, like hurricanes, racial wealth gaps are made worse as relief efforts favor white communities. As these disasters increase in severity and frequency, low-income Black communities and other communities of color become even more vulnerable.[3]

These inequalities are examples of climate injustice, the systematic patterns in which those who are least responsible for the actions causing climate change are disproportionately affected by its impacts. As I argued in chapter 3, negative climate emotions are a part of that, specifically the distressing feelings people of color feel in the wake of storms, heat waves, and fires that are exacerbated by inequality. Communities that have trouble being resilient in the wake of disasters—whether the reasons are financial, social, or due to the general distribution of resources—are also disproportionately vulnerable to mental health challenges.[4] Paying closer attention to climate emotions and mental health in communities of color, including how they shape reproductive plans, will become an increasingly important component of climate justice in the United States.

This chapter shares the narratives of a dozen people of color in their reproductive years. Drawing from a series of in-depth interviews, I prioritize their experiences in their own words by using direct quotes and narratives, and this chapter serves as a qualitative counterpoint to the larger quantitative study presented in chapter 3. The people I interviewed for this chapter were not included in the survey, nor were they drawn from a random group; I started with a small group of people I knew, based on my own location in Southern California and familiarity with the contours of local environmental and climate concerns about wildfire, heat, and air pollution, and they referred others who wanted to participate.

These interviewees have more climate change knowledge than most people do: all of them are college-educated; most of them either grew up or have lived for some time in Southern California; and most have taken environmental studies classes, either as undergrads or in graduate school. They range in age from their early twenties to their late thirties, and none of them are parents, although they differ widely in whether they want children, and whether those who do want them plan to get them through birth, fostering, and/or adoption. What unites this group is the fact that climate emotions are impacting their lives, their plans for the future, and specifically their thoughts and feelings about raising children. As climate emotions researcher Sarah Jaquette Ray argues, the dominant discussions of climate anxiety overwhelmingly center the experiences and concerns of white people, thereby pushing people of color to the margins.[5] This chapter rejects that trend, brings the racial experience of climate emotions into sharp focus, and insists that people of color be heard, here and now.

BROADENING THE FRAME THROUGH FOSTERING OR ADOPTION

Bobby considers himself an environmentalist, and at twenty-two years old, he recently graduated from college in Southern California with a degree in sustainability studies. His family is Guatemalan American. Bobby is both confident that he will become a parent one day and also certain that he won't bring his own biological kids into the world. His thoughts about the environment, the future, and parenting come into sharp relief through his current job, where he is unhappily employed at a restaurant: "There's so much being wasted that could be returned to the earth. That aspect of compost alone is being put into waste and then mixed with toxic chemicals and takes decades to decompose." He connects these waste issues to carbon emissions and how he feels about having children. For Bobby, this is an ethical issue, a reason why he should not have biological children: "With CO_2 emissions, my fear is that, in addition to bringing kids into this world, I'm worried about what they would have to deal with growing up. I was already a young adult when I started to think about these things, but for them, at a young age they're going to have to think about the environment and the fears that come along with pollution. This is why I'm leaning more toward a foster kid, and maybe eventually adopting them. Because it wasn't my choice to have that kid, but I can help guide them to have a better life, versus me bringing a kid into the world who would then have to deal with all of these things. That's kind of my dilemma. The environment is really the deciding factor for me."

Bobby grew up in a close-knit immigrant family where early marriage and children are celebrated, and family is a happy buffer against the challenges of the greater world. His parents and other relatives ask about children all the time, questions he has to duck and dodge. The cultural difference between himself and the older generation is stark: by the time his mother was Bobby's age, she had two children. Meanwhile, Bobby can't see himself becoming a parent before the age of thirty, and he wants to foster a child or children, rather than having them biologically. Although he always wanted to have children, his thoughts about fostering arose from taking environmental studies classes, where he first learned about climate change. It really shifted his perspective, replacing some of his youthful naivete with hard facts: "Going into college was the first time I was exposed to this information first hand, and I realized for the first time, like, wow, it's not all rainbows and sunshine. I had never learned before from any teachers or my family about things like food waste and carbon emissions. That definitely came for the first time in college. And that's when the gears started turning in my head about the future and what I wanted to do, and I started to think this way."

Victoria is the same age as Bobby; she graduated from the same university and is also from an immigrant family, though hers is from Ghana. In Victoria's house there were four siblings and half a dozen cousins who were always around; as a result, Victoria really cherished the closeness and security of a large family. "I guess in the future, I would love to have children," she says. "I'd really like to have a big family. I grew up in a big family, so it's nice." Like Bobby, Victoria is interested in perhaps adopting or fostering, and she also connects the desire for this to her undergraduate education in environmental topics: "I got a degree in sustainability, and I've always questioned bringing people into an environment that's kind of . . . I mean, so much is going on politically, socially, health wise, all of that. I always thought I wanted to give birth, but the more I look at foster care, I realize that I don't need to physically have children to experience being a mom. Like that's not a requirement. I think, in a sense it's a little selfish on my end to think I'm going to have all these kids, when there are already kids in the world who would probably make me a better parent."

Victoria's concerns about biological children are multifaceted: she worries about the future of health-care access, wealth inequality, and whether her children would receive a low-quality education or be racially tracked in public schools.[6] These concerns connect back to climate change for her because they represent a series of challenges that she wouldn't know how to manage with one more thing— like living through a disaster—on top of them. Ultimately, for Victoria, it comes back to how racial inequality interacts with other social challenges to heighten her own sense of vulnerability and that of her potential future children.

"If I have children, they will be Black children," she says. "It isn't self-hatred, I love being Black, but the things I've gone through . . . I wouldn't wish on other children." This is a frequent topic of conversation among Victoria and her friends. They talk about whether they want to have children in the future; most of them do not. Their concerns are quotidian: they are in their early twenties and are concerned about having financial stability and not ruining their bodies. But in a larger sense, they see the world as a frightening place, and racism is a key part of that. "It's scary, and also a big thing is a lot of us met through sustainability and environmental studies, and they're like, 'I don't think the way the world is going . . . I can't imagine bringing children into this world.' One of my friends who has experienced a lot of trauma in this world because of being a Black woman, she doesn't want to bring kids into that."

FEELING VULNERABLE

That feeling, of being traumatized by an awareness of ongoing racial inequality, shaped the perspectives of a group of Black women I spoke to. They were of different ages, from their twenties to their late thirties, and they ranged from just starting out to having established careers. However, each perceived herself, and the prospect of becoming a mother, through the lens of vulnerability. Imagining motherhood as climate change impacts intensify was something that sharpened their already existing concerns to a fine point. For example, Rosalind, older than Victoria at thirty-eight, is a Black woman of Caribbean origin living in Southern California. Unlike Victoria, she is adamant that she does not want children; there is no ambivalence for her. When I ask why not, her answer is simple and direct: "Really, because the world is just awful, and I don't want to bring anybody into it. I just don't think it's fair. Climate change is occurring, but also it's like the crushing racism and anti-Blackness of this world is just . . . yeah, I don't want to do it. I love kids, but, it's just . . . I don't think so."

Rosalind has a graduate degree, a job as a scientific researcher, and is settled in a community she likes. Nevertheless, thoughts of the future are a heavy, ever-present burden. When I ask if there is one issue that feels like the primary reason for not having kids, she answers decisively: racism. She says: "With all of the anti-Black violence, and the police violence against us, I just . . . it just seems so unsafe. And I see so many of my friends who do have children that are constantly stressed because of this, especially the ones who have teenage boys who are taller than average . . . they send their kids out there and then just spend their time worrying about whether their child is going to be targeted or harassed in some way, or potentially killed, and . . . I just don't think I have the disposition to put up with that kind of stress." Rosalind anticipates that the planet, or at least the part of California she lives in, will be a far less healthy place to be, much drier and hotter than in the present. Although she says that her climate emotions are mostly negative, she also finds a sense of hope, even if it is cautious: "Oh, I feel terrible, it's awful [laughs]. It's very sad, and it just seems . . . yeah, it feels . . . it just feels very heavy. And like . . . I wouldn't say hopeless, because there's still hope, but it's less bright than I imagined when I was growing up. Like, I feel like I'll be fine, but the world around me is not fine."

Like Rosalind, Briana, a twenty-three-year-old African American woman, is fairly certain that she won't have children. She lives in the South, and she describes her feelings about becoming a parent as "neutral to negative." This is a generational and age-based issue for Briana; she and her friends feel unstable, like they can't get on their feet. They've struggled to find jobs after college. They've noticed how strained the social safety net is. They worry that being parents would be a difficult, isolating experience. Briana tells me that climate change adds to this through an overall sense of deep uncertainty: "I have been living in places for the past like six years that have been very hurricane prone," she says. "I went to college in Florida. I've been living in New Orleans. Obviously we had a really bad storm season this past August. And every region has their disaster, so it's not fair to say it's only here . . . but it feels sort of unfair to bring a child into this mess. It's a multifaceted mess. But climate change is pretty high on the list of reasons [for not having a child]."

Briana finds it difficult to make long-term plans, particularly when it comes to where she'll live; planning a future in a place with severe storms that are increasing in frequency doesn't seem to make sense. All of this has a clear emotional impact as well; when I ask about the emotions she feels in response to climate change, she says: "Anxiety . . . I wouldn't say I feel hopeless, because there is change, there are changes that can be made around climate change. But I think anxiousness for the future for how it's going to be handled by those in power. Bleak, sometimes the future feels bleak."

Nia is African American; a thirty-three-year old Millennial from Georgia, she works as a researcher with a focus on air quality. Unlike Rosalind and Briana, she is undecided about whether she will have children in the future. Her stable job gives her a financial foundation, but environmental issues weigh heavily; climate risks for Nia are closely tied to racial inequality. "Race politics make me feel deathly afraid to have a child, even more so than climate change," she says. "Because with climate change, at least if everyone's burning up, they'll do something about it. But we have a long history of people ignoring the problems that hurt people of color the most, which is why environmental justice exists. So as long as climate impacts are worse in communities of color, it will be harder for us to adapt and be resilient. I think it is cruel to birth a Black child in the United States."

INTERNAL DESIRE, EXTERNAL CHAOS

Not all of the people I spoke to were uncertain or negative about having biological children. Martina, a twenty-seven-year-old Black woman in Baltimore, is a public school teacher, certain that she wants to be pregnant and birth a child. At the same time, climate change is an emotional experience for her. "Sometimes you walk outside and you're just like, 'it should not feel like this at this time of year' . . . like, the weather, or the climate I guess, is changing my mood because it's way hotter than it's supposed to be, way more of the time. So I can already see how the climate affects my emotions, and I could see how that could get worse as climate change gets worse." Martina's climate emotions mostly feel minor and temporary, though, and she is careful not to remain in a negative emotional place. "I think emotions are temporary and nothing is permanent, but you have to be aware of those things . . . the world around you can really affect the way you feel inside, and you have to be mindful about that, because if not, you could easily get stuck in those negative emotions."

These feelings don't easily map onto feelings about parenting. Martina wants to have children in the future, although she's ambivalent about it and feels unprepared. Her main concerns about it are not necessarily about climate change itself, but rather her own mental state and whether she would be able to manage children from a mental health perspective. Climate change is something that would map onto that already existing mental state: "Thinking about climate issues and the emotions that come with that, when it gets to the point when I feel ready to have children, I feel like I would have to be really mindful of those emotions and how I can manage them. *If* I can manage them." Having an awareness of the challenges of pregnancy, birth, and raising children is not a barrier to wanting that experience per se; neither is climate change. Instead, they are multipliers that compound Martina's perceived sense of vulnerability and uncertainty in the world.

The same is true for Melanie, a twenty-six-year-old Native American woman raised on the Navajo reservation and in Southern California. She idealizes having a big, happy family, but there are aspects of the world that give her pause, so she struggles with whether it's morally okay to have children. "I think, just kind of like morally, I think I may not have children although I do want them," she notes. "Just because, with all of the things we see going on in the world, it seems unfair to bring someone into all of this against their will. And it's like, good luck, deal with what we left you! I have no idea what the world is even going to be like when my child gets to be my age, and I don't want them to have to deal with that. That's really scary." Melanie's feelings about climate change include a general sense of powerlessness and lack of control over other people's actions, which directly translates into her fears about parenthood: "With climate change, we're the driving force of things breaking down but then also, the planet's going to do what the planet's going to do and we have no choice in the matter. So that can be really scary as well. And then it feels like . . . and I don't know if this is all generations or just my generation, but it almost feels like, kind of shameful to want to have children. People look at you like, *eww*, why would you want to bring children into this world?"

THE NEW PEER PRESSURE: DON'T HAVE KIDS

I was surprised to hear from several of the people I interviewed, as well as anecdotally from a number of my students, that there is negative peer pressure about wanting children, particularly among younger Millennials and those in Generation Z. "If you have friends who are socially conscious or environmentalists or anything like that, then it's kind of like shamed to want to have children," Melanie states. "Because you're bringing another person into a world that's already pretty crazy, and you don't know where it's gonna be years from now, like why would you do that?" Like her peers, Melanie is concerned about her future children and their potential vulnerability for a variety of reasons—climate crisis is just one. "I don't know what my child is gonna look like, I don't know if they're gonna be straight, gay, or if they'll have a disability, all of these things. And . . . their life could possibly be ten times harder, or their life could be put in danger for simply existing. So there's those factors as well on top of the climate thing. There are just too many things to be worried about, all at once."

Juliana, a twenty-three-year-old Mexican American woman, is strongly aware of negative peer pressure from friends. She recently graduated from art school, and her friend circle is mainly composed of queer and transgender, antiestablishment artists. Most of them have no intention of having children of their own, which seeps into conversations with Juliana. She says that "a lot of them don't want to have kids. A lot of them are very anti that. Um . . . yeah, I'm hopeful, but my friends aren't coming from that perspective." Her friends cite environmental concerns and mental health concerns; their anxiety tells them that they can't properly take care of themselves, much less a child. They also struggle, as trans and nonbinary people, with the issues of access to fertility centers and the need to use reproductive technologies that feel out of reach.[7] For both environmental and social equity reasons, Juliana feels that it may be unfair for her to consider having biological children. She tells me that these feelings are not entirely separate from how she feels about what her child's racial upbringing would be. As a dark-skinned Mexican woman, she regularly experienced racism growing up in Southern California— and given that her husband is white, any child she might birth would be biracial, which raises questions about whether and how they would navigate the world differently than she has.

However, Juliana is an optimist; she does plan to have one child for sure. And because she is currently a school teacher, she feels that she has some idea of what to expect when it comes to children and climate change: "I notice it a lot with my students, high schoolers and middle schoolers. They are sensitive . . . *very* sensitive. I don't really get into it with them, but I know that they're dealing with con- stant anxiety. When they talk about climate change, that's been a conversation. They don't want to drive, because they know that's another just car that's contributing to pollution. Or with meat, they'll say the cows produce emissions that are killing our planet. And . . . so I think there's that."

IS KNOWLEDGE ALWAYS POWER?

In some circles there is a common assumption that climate action, specifically activism, is a form of antidote to climate distress. Cindy challenges that perspective. At twenty-one years old, she identifies as a queer, mixed-race, Asian American woman, and a climate activist on her college campus. Unexpectedly, the first environmental emotion Cindy identified with was embarrassment: "Yeah, I think when I first started learning about environmental justice in general, like my immediate reaction was embarrassment, actually. Because, I mean, I grew up in the US, and like, these things aren't affecting me nearly the same way they are affecting other people across the globe. And learning about that . . . I was just like, *I am the problem. What can I do to stop this?* And, you know, that embarrassment soon turned into frustration and anger, like why aren't people addressing this on a global scale? Yeah, the two main emotions for me were definitely embarrassment and anger. But there was also frustration and also hopelessness as well."

The embarrassment was triggered by the common climate problem Californians deal with: wildfire. The 2018 Woolsey Fire broke out when Cindy was in high school and provoked a lot of climate anxiety for her, which led to her feeling embarrassed: "The Woolsey wildfire . . . it tied into a lot of the fear that I was feeling. We had school days canceled completely. We couldn't show up to school because of the air quality. And the day that we went back to school, you could literally see like ashes falling from the sky. I grew up in L.A., so I was like, *Wow, is this like what snow days feel like? Because this is absurd.* And so I think a lot of that embarrassment also came from the fact that I knew these things were actually happening to other people at a much greater scale than it was to me and much more frequently."

I ask Cindy if she ever wonders whether she'll be a parent in the future. She tells me that it is a constant thought. Although she is very young, she grew up with religious parents, and she has felt pressure around planning for future children. However, climate change, and her years-long campus activism, led her to question having kids based on moral grounds. "Can I morally justify bringing a child into this world if I myself came into this world, saw the havoc created by climate change, and I know it's just going to be worse?" she asks. "I feel like I don't want to be responsible for all the things that have to come with existing in a world with so much inequality and like environmental injustice in general, it doesn't feel fair for me to have a child in general, I guess."

Cindy's concerns aren't solely about climate; they are also about racism, sexism, and the inequalities in the world in general. My interview with Cindy ended on a hopeful note, though. The future was something she actively avoided thinking about for a long time, but now her own commitment to climate activism, and that of others she works with, makes her feel hopeful: "I started working with people who are leaders in climate change, or climate action. And it gave me a lot more hope, and made me realize that there are things that people are doing right now to address climate change at a local level. I guess thinking about coalition building within climate justice, and also creating networks of solidarity in general made me a lot more hopeful. . . . I try to remind myself that, you know, people are doing the work, I'm doing the work. And it's something that we just have to keep pushing forward."

Luz, a twenty-four-year-old mixed-race Latina who is currently in college, is also active on environmental issues on her college campus. She experienced an emotional reaction to climate change for the first time in an environmental studies classroom; despite the urgency of the challenges discussed in the class, Luz found that it validated her feelings and sparked a passion to take more committed climate actions. It helped her draw connections between the climate future and the present—specifically the environmental and health issues already plaguing the planet, how they might impact future children, and how this might impact her current relationship.

She says: "There's already a lot depending on my own generation to fix stuff, and I don't know what would be expected of my kid's generation. Because I don't see these issues getting better. And I also think of . . . even just health wise, I visited L.A. just a few months ago, and I felt horrible there. Just with the difference in smog and air quality there compared to here [Santa Barbara]. And I know that it's a real thing that causes health problems, and I don't know what kind of effect that would have on my own children. I was talking with my boyfriend and told him that this interview was coming up, and he asked, 'So what do you think about having kids?' And I raised the issue of adoption. I feel like it's selfish to have your own kids when there are so many kids who need a home and a family." In other words, climate change is not only remaking how Luz thinks about children, and whether and how to raise them, but also the romantic relationship that might get them there.

HOW DO WE DATE NOW?

I spoke to several young women who are addressing the kid question with their dates, potential partners, and long-term boyfriends. At twenty-two, Elena is one of the most certain people I've met: she is not having children. She's from a Salvadoran immigrant family in which she is one of four children, while her mother was one of twelve. Given that Elena comes from a large family, her certainty that she won't have children is a surprise. She responds that it stems from both life experiences and climate fears: "Me being interested in environmental policy, that cemented my decision to not have kids, but I do have some personal things that I've gone through in life that I wouldn't want my kids going through, like not having a dad. So I feel like it's best if I just focus on myself and take care of my mom. Like, I don't think I should be bringing a child into this world . . . not only is the planet doomed, I can also like, spend my time and energy focusing on someone that's already here. I don't have to . . . I don't want to bring a child into this world. I just don't see myself being a parent."

Elena brings this conversation up on the first date with any new guy she sees. Given that most of them expect to have families in the future, Elena feels strongly that she does not want a relationship. This has been discouraging for her, but her mind is made up; she is adamant that she will not meet someone, fall in love, and change her thinking. She is childfree by choice, and she plans to stay that way; as a result, Elena's future plans include travel, spending time with her mother and other family members, and not becoming anyone's mother. Like Cindy and Luz, Elena's feelings about climate change were sparked by environmental studies classes, which got her thinking about how long-term environmental problems would diminish her future children's quality of life.

She says: "[I] started feeling like having kids is definitely not a sustainable thing to do. Like, I had my personal reasons but I don't want them to grow up and ever have to leave their home because of sea level rise. Or have to be worried because of really weird weather patterns. I wouldn't want them to have to go through that, on top of whatever mental health issues they may have. Gen Z is already going through stuff, just imagine what those coming after Gen Z are going to have to deal with. They're just going to get the worst of it . . . so it's just like, I'm very pessimistic. I know that things aren't going to get better. So why would I want to put a child through that? Even when my sister gave birth to my nephew, I was like, *Why? They're gonna go through so much.*"

Climate-driven kid questions shape relationships and dating for several of the people I interviewed, but not necessarily in the direction of not becoming parents. Elena's close friend Veronica, a twenty-two-year-old from Los Angeles, manages the cultural expectations of a large, immigrant family from Guatemala, where she often feels she's expected to already start having children: "Because of my Hispanic background people are always like, *when are you gonna have children, of course you're having children.* It is what it is, right? But now that I'm like, an adult, I think about it differently . . . would my child have a good quality life? Will they be able to survive? I feel like I have cultural pressure, but I also want to be mindful of that child. Because it's not just about having it, it's about raising it. And being able to sustain it as well."

Veronica does want children, and she plans to have them someday with her boyfriend, who she has been in a relationship with for five years. They've discussed the issue, and he wants to move forward with creating a family soon. However, she's hesitant—about his family, their values, and the everyday realities of environmental toxic pollution in the air and in people's bodies. For Veronica the everyday environmental concerns link directly to the larger issues shaping climate change: power, who has it, and who doesn't. Though seemingly distant, intergenerational power imbalances—and older generations' legacies of generating the emissions that have caused climate change—make her feel that it is unfair for people her age to have to ask the kid question. She says: "I just think that people in power, whether they believe in climate change or not, it's not beneficial for them to really do something about it. Because they're older, it's not going to affect them the way it affects us. So the deniers are right: it doesn't exist for them. Not the way it exists for us. Because they have so much money and power that it doesn't affect them the same way. It doesn't hurt their lifestyle. They can buy protection from what the rest of us are going to have to deal with."

WHY WE STILL WANT KIDS

Although this chapter has focused primarily on the challenges young people face as they approach reproductive questions, many of those I spoke to still wanted families of their own. For those who were certain about having children, the reasons were emotional: love, joy, happiness, and hope. For example, Bobby was clear that he doesn't plan on having biological children, but he was happy about the thought of fostering in the future and was particularly excited at the thought of his sister having kids. "I would love to be an uncle," he said. "I love the niece I already have; she's been coming over more and more lately. We go on the swings, play games, read. Just seeing the next generation, the reason why I've been more optimistic about having a foster child of my own, is about being able to see them grow. To see my family push forward even though we've also had negative

things happen too. I'm excited for the future with my sister, specifically her future family."

Victoria was excited at the prospect of adopting multiple children,
to re-create the happiness of the large family she grew up in. "I always
have siblings to talk to or someone to connect with who really knows
me and my home life in that unique way, and I've always wanted to
create that for someone else," she noted, smiling. "I want to create a
space where kids have loving, supportive parents. My parents aren't
perfect, but I know that I grew up in a loving home where they would
do anything for my success and protection, and I want to create that
for someone else." Her sentiments were echoed by Melanie, whose
experience living in a racially and gender-diverse family inspires her
to want to re-create the same. She said: "When I look within my own
family, we're very diverse. We're Black, we're White, we're Native
American. We're straight, we're queer, we're nonbinary. And we still
have compassion for each other and that kind of spills over into compassion for other people that we don't know. And I think, like, I don't
want to quit. I don't want to let the bad things dictate how I make my
decisions, and, you know, if every good person decided they didn't
want to have kids and bring them into a terrible world, then there
would only be terrible people left. The idea of bringing someone into
this world and growing them with compassion and love, and making
sure they grow up knowing to stand up for other people and stand up
for what's right, that's a little glimmer of hope."

Clearly there is no single way of navigating climate emotions and
the kid question. Race, gender, sexual orientation, class, culture,
experiences of discrimination and privilege—all of these factors
position people differently with respect to their climate emotions
and, consequently, how they think about having children, both now
and in the future. However, for the people I interviewed, a general
awareness of their own existing social vulnerabilities, as well as how
they might be compounded by climate emotions and direct climate
impacts, gives them pause. This is the unjust element of climate
anxiety: the fact that this anxiety is layered on top of structural and

systemic inequalities, fears and concerns that future children will endure discrimination, injustice, and an increasingly unlivable planet—and that their parents will not have the ability to buffer that and protect their children. From my perspective this is even more of a reason to fight to advance climate justice and to recognize that emotional, mental, and reproductive justice are key components of that fight.

5 The Personal Is Political

It's Saturday morning in August 2021, and I have a Zoom meeting to attend. I'm not exactly thrilled. A year and a half into the pandemic, most of my socializing still takes place online. As much as I appreciate that Zoom has kept my social and professional life afloat, I'm really tired of it, which is why I'm not looking forward to what would otherwise be a fascinating event.

I'm here for a virtual house party. Not the kind of house party you're probably imagining, but rather one that's designed to talk about climate change, the emotions it brings up, and how these emotions are disrupting our collective confidence in being able to have kids. In other words, reproductive anxiety and the kid question. This party is the brainchild of Josephine Ferorelli and Meghan Kallman, two Millennial climate activists who have been gathering similar groups in person in the Midwest and Northeast since 2015. Working under the banner of their nonprofit organization, Conceivable Future, their aim is to raise awareness about the threats posed by climate change—specifically reproductive threats—and to use that awareness to advocate for policy actions such as ending fossil fuel subsidies. I had been researching Conceivable Future's work since 2019 and was excited at the prospect of learning what a virtual house party would be like, the first of its kind since Covid-19 had forced us onto Zoom. I was not, however, excited about being an actual participant—sharing my personal perspectives on the kid question.

As a researcher, I've come to expect a certain amount of distance from my topic. I've been talking to people about the subject for quite some time in the mostly one-way dialogue of the research interview and have honed a style that is approachable, that coaxes people into expressing some of their most intimate feelings about climate change, relationships, sex, the future, and the possibility of ever parenting. Some interviewees weren't accustomed to talking about these subjects openly, not even with their spouses or families; I was used to getting people comfortable with having the conversation, but it was never a dialogue because I never expressed my own feelings.

That is intentional—I maintain some aspect of distance from the subject at hand, at least during interviews, so that I don't unduly influence the person I'm talking to as the researcher. If I'm focusing on my own climate anxieties, or the ways I wish I could bury my head in the sand and pretend that it would all just go away, that might sway the person I'm talking to so that they skew their comments in the direction of my own. But if I'm honest, I also hide behind the researcher role because it's simply easier than telling the truth—which is that I have a lot of feelings about climate change, and they have little to do with scientific studies and projection models. They have to do with everyday life.

Concerned, scared, worried. That's how I feel about climate change. I was *concerned* when, after finally purchasing my first home, I had a hard time buying a hazard insurance policy to cover fire damage. Few insurance agencies are willing to provide fire coverage for people living in my new neighborhood, based on their algorithmic analyses of how much dry brush is in the area, where that brush is located relative to my home, and—given how arid the region is—how likely it is that a wildfire would pose a threat. I was *scared* when, several months after moving in, I looked through my bedroom window at what seemed to be a dust cloud approaching from the direction of the mountains in the distance, only to find moments later that it was smoke from two small local wildfires, no more than three miles away.

I put on my KN95 mask and walked down the street to get the mail, eyes burning with every step, passing a neighbor along the way. We shrugged at each other as if to say, "What can you do?," but inside I was shaking. What could I do if the wind whipped up and made the fires worse? What if that happened in the middle of the night? I was *worried* when the temperature hovered around a hundred degrees throughout the summer as a national heat dome baked the entire country, turning the local vegetation in my area into a tinderbox and sending air pollution readings sky high. Even a brief break from the worry, when I traveled north to the Bay Area to be filmed in a documentary, was tempered by my friends' tales of the apocalyptic, orange sky–inducing wildfires that had hit the area the previous fall. *Concerned, scared, worried.* Most of the time, my feelings about climate emotions have nothing to do with my thoughts about children. On occasion they do, mostly in the brief, fleeting thought: *I'm so glad I don't have a kid to worry about on top of all of this.* But this is all personal, right? Who else cares about my feelings anyway?

In this chapter I move back and forth between self-reflection and tracking the growth and development of an advocacy organization that formalizes conversations about the kid question. My aim is to understand the ways race and gender dynamics intertwine to shape Conceivable Future's work and how it is perceived by its founders and broader circles beyond them. While the organization's work is a form of gender-based climate activism, it is situated alongside long-standing ideas that elevate white women's childbearing into an issue of national importance. It was not Meghan and Josephine's intention to harness those prevailing narratives. When they launched the organization, they found out that public acts of reproductive resistance, such as launching an organization questioning the viability of having children during the climate crisis, are a lot more complicated than they anticipated.

Yet their experience raises key questions: Whose reproductive anxieties come to matter in public, and how? What difference can be made by bringing these "private" concerns into public spaces? And, even when the stakes are quite different across race and class divides, what can making space for conversations about climate-driven reproductive anxiety offer to a broader range of people?

PERSONAL EMOTIONS, POLITICAL IMPACTS

Conceivable Future emerged from a belief that the founders' personal emotions about climate change and the kid question were publicly relevant, even important to other people their age. From their first meeting at a very different kind of house party in late 2014, Josephine and Meghan began an intense conversation about climate change, their feelings about motherhood, and how they felt that the fossil fuel corporations responsible for the climate crisis were foreclosing their ability to realistically imagine bringing kids into the world. When I interviewed Josephine and Meghan, they both described their initial meeting as feeling intense, significant from the start.

"The feeling I remember most when coming together with Meghan was excitement and relief, like a light came on," Josephine said. Growing up in the nineties, an era of intensifying global warming and political inaction, she had become accustomed to dealing with her feelings about climate change in private, even questioning whether they were real or significant. "There was both this sense that it was terrifying, but also this confusing thing of maybe it wasn't really happening or maybe it didn't matter," she said. "I was always looking for evidence that it either was or was not serious, or that it

was or was not happening. So when we met it was this sense of relief."
Josephine was also dealing with the acute grief of having recently lost
her father. When she and Meghan came together to discuss these
issues, those emotions were at the forefront of her mind and heart.

Meghan was excited by how their conversation went straight to
the heart of some of her thinking about gender and climate activism.
She had recently been involved in supporting a public action by two
male friends who had used a lobster fishing boat to block a coal ship.
She was concerned about the ways their successes in that action
were shaped by the hero narratives that often uplift male activists.
From the beginning, it was clear to her that the ability to engage in
public climate activism is gendered: "No woman I know would be
able to do an action like that and have this kind of John Wayne experience where the D.A. would drop the charges and then everyone
would go to the People's Climate March and hold hands. It was an
incredibly masculinized experience of activism. And I'm supportive
of that action, I know that direct action like that takes a lot of cour-
age and organizing. But I was feeling sort of raw about the gendered
patterns of [it]."

Unlike me, Josephine and Meghan did not assume that their personal feelings were theirs alone; rather, they started from the premise
that these were likely shared feelings experienced by other people of
reproductive age, and that they could serve as the basis for collective
political action. They began brainstorming what those actions could
be, discussing everything from a "baby blockade" to a sex strike, and
finally decided on private house parties. These would be spaces
where emotions would be shared, collectivized, and translated into
political action through recorded testimonies.[1] In other words, they
created a strategy focused on making the personal political.

"The personal is political" was an important maxim in second-
wave feminist circles in the 1960s, connecting the everyday experiences of women's personal lives to broader social, political, and
cultural forces. It was first described in writing by feminist activist
Carol Hanisch, whose text "The Personal Is Political" analyzed the

role of women's therapy groups. This strategy grew in popularity as the women's movement gained momentum. These were small, analytical discussion groups where participants analyzed the everyday challenges of daily interactions in women's lives, homes, and relationships, working to find collective solutions. The groups were not focused on identifying and addressing individual people's "personal" or "private" problems to offer individual solutions; instead, they uncovered and addressed how these so-called personal concerns were connected to larger patterns of gender-based inequality. As Hanisch noted, "there are no personal solutions at this time. There is only collective action for a collective solution."[2]

From the beginning, Conceivable Future was conceptualized along similar lines. Meghan and Josephine decided to organize house parties and recorded testimonies as the vehicle for sharing emotions around climate and reproduction based on the power of their emotional appeal as well as the ability to connect that appeal to broader circumstances facing others. As Josephine described it: "Someone is telling you something that is really deep to them, and it might help you connect to your own stuff, it might shock you into awareness. But their greatest function is for the people who make them and the communities they're in. If you make a testimony, you're making yourself vulnerable and rising up to an occasion by sharing something meaningful to you. . . . It matters less whether a million strangers see it than if your mom sees it. Or your boss, your legislator, your siblings, whoever. . . . It's a big risk. And there's a lot of vulnerability there." They created a website advertising this approach, with a mission stating that "the climate crisis is a reproductive crisis. . . . We demand the right to make reproductive decisions free from massive, avoidable, government-supported harm."

As of spring 2023, there are more than ninety testimonies on the site in written and video form.[3] They reveal a broad range of emotions and ways of thinking about climate change, reproductive possibilities, and the future, and they were created primarily by young white women who participated in the in-person version of the house party I attended online. Whereas people of color tend to connect reproductive anxieties to larger social and structural issues including racism, class inequalities, and broader fears about the kinds of discrimination their potential future children would face in addition to climate change, testimonies from white house party participants are more narrowly focused on individual feelings, including moral and ethical concerns about what it means to be a "good environmentalist." These concerns are important, but not because they represent a broader or universal experience. Rather, they are important because they reveal some aspects of how race and class privileges shape a very specific experience with the kid question.

For example, a twenty-three-year-old woman from Indiana spoke of her fear and sadness that the family she's always imagined would never come about. "So . . . ever since I can remember . . . I've wanted to have redheaded babies," she said. "And my family has wanted me to have little redheaded babies . . . however, as I've learned more about the state of the world today, I've become more and more skeptical about whether or not it is safe or ethical for me to bring more little baby humans into existence. With an increasingly changing climate that is just increasingly plagued with more intense and unpredictable natural disasters and, you know, I question what kind of life my little redheaded babies would have. I question what kind of future they would have and I'm afraid. I'm afraid that they will have to live in a world that is increasingly full of war fueled by climate change. But I am also deeply afraid of not giving my parents the beautiful little redheaded babies that they have always wanted."

Her feelings were echoed by a thirty-one-year-old woman from Rhode Island who, despite wanting children, was deeply conflicted and afraid at the prospect. "So I've been conflicted about having children," she said, "and one of the main reasons that I've been conflicted about having children is the question of what kind of world I'm gonna will to that child, right? I have a niece, she turned two last week. And when she was born I remember having these conflicting emotions. I was so happy and I was terrified at the same time. I mean, I remember holding her at three days old and it was one of those—she was born in August, and it was 105 degrees or something insane like that in northern Vermont where it's not supposed to get to be 105 degrees, right? And it was this sense that the things that we depend on are now undependable because we've messed with the systems that kept them dependable for so long. It is also difficult to have one's reproductive future conditioned by people who don't care about you."

A third testimony, given by a twenty-three-year-old from New Hampshire, was focused on the moral question of "fairness"—whether it is "fair" to bring children into a world already facing so many challenges, and whether those future children might contribute to those challenges. "I've thought about this in the context of two questions," she said. "The first is: is it fair to my child to bring them into this world? And then, is it fair to this world to give them my child? First because when I talk about having a child in a world that feels so hopeless, where there's going to be a lot more disease and a lot more natural disaster and a lot of those issues, is it fair for a child who's going to have to face those crises and who's going to have a quality of life that is worse than mine in those ways? But I think that coming from a place of privilege in a wide range of things—being American and college-educated and white and living in a place in New England where there are fewer climate impacts than in many other parts of the country and in the world and where I'm not constantly exposed to contaminants, I think that . . . I am much more concerned by the second question. I'm much more concerned about what my child might do to the world and what kind of resources that they'll use and the way that they will be contributing to those problems."

Finally, a twenty-one-year-old woman in Canada spoke about her feelings of anxiety, and the surprise that came along with suddenly questioning for the first time whether the family she wants is a possibility. "I was struck to hear how many other people are also feeling the same anxieties as me about whether or not we should be having kids in this current context, which is obviously a very personal question. But myself, I had always wanted to be a mother since I was little. It's just something I've kind of never kind of questioned and I always wanted to do. And then about a year ago I got more involved with climate justice activism, and a few months ago I started questioning for the first time whether I should still be having kids right now just given the world, the direction that we're headed . . . and I know that many of my friends, especially those involved in climate justice activism, are wondering the same questions as me and some of them are planning on not having kids. And this is a big reason why . . . just because it feels that our choices are severely limited."

Clearly these young women are asking the kid question in many of the same ways as the people I interviewed in chapter 4. However, their concerns are shaped by something that seems to be new for them: the possibility that the future that's coming will be unsafe and that their government leaders and other decisionmakers will not protect them.[4] Instead, policymakers are actively making decisions that will make their lives, futures, and children unsafe through increasing climate impacts. While this is nothing new for people who have experienced the structural violence of racism, poverty, sexuality, or non-normative gender identity, it is a shocking realization for those who are confronted with it for the first time. And, like Meghan and Josephine, it drives them to want to collectivize their concerns.

WHOSE REPRODUCTION? WHOSE POWER?

Before cofounding Conceivable Future, Meghan had spent a lot of time thinking about gender and climate activism, and how reproduction could be a powerful way of drawing the two together in a public conversation. She was familiar with the history of reproductive politics in America, in which women's capacities to become pregnant or not, sustain a pregnancy safely, deliver a baby, and raise a child with the necessary resources in place, have long been regulated by powerful political and medical institutions. This history was personal for her, as she grew up in a low-income family and was well aware of the narratives stigmatizing the reproduction of the poor. At the same time, having terminated a pregnancy of her own, she came to understand that her reproductive choices had power—potentially more power than the more traditional climate organizing she had been involved in. Reflecting on the experience later, she wrote: "This was what I had to politicize in order to talk about climate and injustice. This was where the power was. The realization was like a cold stone gradually settling somewhere in my middle, the weight of which has been present since."[5]

Women who gave recorded testimonies noted the powerful consequences of their reproductive decisions, in terms of broader significance to communities far beyond their own. One twenty-three-year-old said: "Now we're dealing with this whole other existential crisis that in the past like never, we never had to think about. People like either wanted to have kids, or they were forced to have kids, or they didn't. And now it's not just about ourselves, it's about the state of the world and how this decision doesn't just affect us but affects all these other people and the way our children are going to be in a place that is maybe unsuitable for humans or full of crisis . . . and that's just something that's completely new and really, I think more than maybe anything else brings to light like what the climate crisis means."

A thirty-two-year-old from Chicago grounded the issue in the power of not reproducing specifically through a human rights lens. She noted: "In my early twenties I really began questioning whether I wanted to have kids, and had mostly been leaning no for a number of reasons that were personal, professional and economic as well as environmental. But now I think facing facts of climate change, how close we are to two degrees, how uninhabitable the places where we grew up are going to be by the time we die. I think the decision to not reproduce has become an increasingly easy one for me to make. . . . I see climate change as a human rights issue. As something that is everybody's problem. And as something that is not only not isolated, but actually intersects with and *compounds* basically every other problem that people care about. . . . I guess the last thing I would say is that my decision to not reproduce comes mostly from a place of love. I'm really concerned for my friends, and families, and people I've never met who have kids or are about to have kids. I'm concerned for the struggles that they will experience in an increasingly hostile environment."

Historically in the United States, white women's reproduction has been publicly linked to the health and vitality of the nation. Reproductive rights campaigns led by white, middle-class women have typically focused on accessing abortion and contraceptives, ways of not having unwanted children, while—for women of color— this has not always been the primary concern. Even as middle-class white motherhood has long been encouraged and promoted, at times even coercively, through social policies and public campaigns, women of color and poor white women have battled population control campaigns focused on reducing their numbers in the broader population.[6] These parallels are stark within environmental activism: when white environmentalists began to encourage white women to have fewer children in the mid-twentieth century, their campaigns focused on personal responsibility, moral leadership as environmental stewards, and the freedom to adopt social roles outside of motherhood. Meanwhile, women and girls of color were being subjected to widespread coercive reproductive policies and practices, including sterilization.[7] Prominent American environmentalists were careful to avoid race in their messages about population and reproduction in the United States, or if they did address it, they focused on reducing the consumption practices of the white middle class. While they sought to avoid racial controversy and accusations of population control from people of color, they effectively established the idea of having fewer children as an environmentally and morally responsible act as the domain of young white middle-class people.[8]

These ideas show up in current-day testimonies as well. A thirty-one-year-old white woman from Vermont connected population issues to a sense of personal responsibility; for her, being responsible meant not having children, to balance out personal desires with impacts on the planet. As she put it: "As I became more and more aware of environmental issues, natural resource constraints and climate change, and the connection to population, it started to really strike me that there's like, a really . . . strong connection between, especially the number of people in the first world who are contributing to climate change. And then also, the consequences that the ecological crisis is going to have for human society. So those two factors are really the biggest concerns for me. . . . So, there's just this part of me that feels like it would be really hard to know that that was going to be my children's future. . . . I still continue to be really conflicted about these issues and so, um, it ends up being this sort of painful, personal dialogue in my head around like the global versus the personal. And like the common good versus what I want out of my life."

Anyone with a passing familiarity with these issues knows that reproductive politics is a polarizing subject. Many white environmentalists advocate population reduction interventions, viewing them as necessary for sustainability and resource conservation; for many people of color, the history of population control has always been a racist history and supporting population reduction cannot be separated from that terrible history. Regardless of one's position, raising questions about the viability of having children—particularly in the context of environmental crisis—is often interpreted as an implicit moral or political statement about whether people *should* or *should not* have children and exactly *who* should have fewer children.

In this context, Conceivable Future has generated no shortage of controversy. Soon after the organization launched, journalists began to write about it, inaccurately describing Meghan and Josephine's work as encouraging young people not to have children. But behind the scenes they had had long, careful conversations with a range of people before creating the organization. They had asked friends for feedback on how to approach their work and were advised to simply ask and listen, rather than trying to tell anyone what to do. Armed with this feedback, they "rushed into the void" by creating their website. But the strategy backfired. Media reports, social media commentary, and conversations with acquaintances let them know that the conversation beyond their circle was nowhere near as nuanced as it was inside of their personal community. Now Meghan and Josephine were engaging with the broader public, and the public saw population as a clear driver of climate change. The public comments in articles written about their work were filled with troublesome racial and class dogma assuming that women of color and low-income people don't know what that right answer is for themselves; commenters offered suggestions focused on controlling women's bodies first rather than finding ways to change the broader systems that constrain them.

It is true that there was a lack of racial awareness in Meghan and Josephine's early approach, which translated into how media and the public have taken up their message, and in some cases, misunderstood it. They admit that they were naïve at the start of this work; they weren't particularly thinking about how race and privilege frame their reproductive experiences, mainly because their own privilege shielded them from having to do so. While Josephine concedes their naïveté, she argues that their original intentions were broadly misunderstood: "Our original aim was, how do we get people to acknowledge the emotional stakes, to stake their own claim to the climate crisis, and to move from individual, paralyzing guilt and shame, to a place where you can hold that and still take meaningful action? What happened when we stepped into proposing the idea of reproductive lives in the context of climate crisis was that people had been conditioned to think of this only in terms of population control, and so they thought we must be offering the so-called solution to the climate crisis, of telling people not to have babies."

When Meghan and Josephine first became aware of reproductive justice, they understood it to represent a framework for addressing how people's reproductive capacities are shaped and constrained by broader political and societal forces, and how the context of having or not having children is, particularly for many women, not a simple matter of making private choices. They saw this as a meaningful way of articulating their own concerns, despite the fact that they were not addressing the most important aspect of the approach: resistance to reproductive oppression—which, for many women of color, stems from a long history of reproductive racism in the United States. So they added references to reproductive justice to their website and began talking about it in interviews. This brought resistance of a different kind—from long-term reproductive justice advocates, who critiqued their lack of analysis of race and class oppression.

In response, Meghan and Josephine conducted more research and shifted their approach a bit. Reflecting on this shift, Josephine remarked: "Over the years I've become more aware of the appropriative nature of how I first engaged with it. I don't think we're good ambassadors for reproductive justice as a movement, but we're just people who have found a lot of meaning in that framework and who are trying to share it with people who hadn't yet encountered it." Meghan agreed, noting that they have had many conversations about their own whiteness and whether they should be known as a reproductive justice organization: "Again, it's one of those things I feel conflicted about. The tool is very useful, but Josephine and I have said to each other, 'We can't be the reproductive justice people. We're white.'"

Nowadays, when journalists make inquiries about Conceivable Future's reproductive justice work, they redirect them to the well-known SisterSong Women of Color Reproductive Justice Collective. However, the stated mission at Conceivable Future still describes the climate crisis as a reproductive justice crisis, without addressing the racial dimensions of this way of framing the issues. As far as Meghan and Josephine are concerned, though, this is their imperfect response to a long process of asking themselves how to best engage a framework that has offered them a lot of richness. They are aware that using the term helps spread the message and expand the analysis, but as white women without a racial analysis of climate change and how it shapes their reproductive possibilities, they are not ambassadors of the movement. Meanwhile, many of the people who are drawn to the website, the house parties, and the testimonies are white, and racial power dynamics go unremarked in many of their testimonies. This is particularly the case among those who focus on population growth as a main driver of climate change.

For Meghan and Josephine the most important thing is that they are working to chart a different course, one that helps people to move away from scrutinizing individual reproductive choices while building a "movement of radicalized people who can claim their own stake in the fight and use personal pain to build moral power."[9] While babies are symbolic of the fight, the fight is not actually about the babies. It is about the systemic and institutional forces that are driving a future in which babies and families don't seem like a viable possibility for people who want them. Articulating this painful reality through emotional testimonies has a particular power rooted in truth telling, even if those truths are offered through a narrow lens.

HOW POLITICAL IS MY PERSONAL?

Finally, I log onto Zoom and join the house party. We settle in and do a round of introductions, followed by the facilitator giving an overview of how the meeting will go: begin with background explanations of the house party approach, testimonies, and Conceivable Future's mission; establish shared ground rules for participating in the day's activities; learn how to prepare and record a personal testimony if we want to do so; a free-write session sparked by a prompt, which we will share with partners in small groups and again in the larger group. I listen and nod, wondering about the others in the meeting. There are ten of us; the others are mostly in their twenties and thirties, mostly white, with perhaps two or three other people of color. I wonder whether their minds are already made up, whether they're here to judge or tell others that having children in the climate crisis is selfish and dangerous. I wonder if I'll be able to relate to anyone in the group, given that so many of my own feelings about climate and reproductive anxiety are also about racial vulnerability. I'm wary that someone will bring up population issues; I don't feel like having that conversation today.

For the free-write session, we're prompted with the question, How is climate change shaping your reproductive life? We're given ten minutes to respond. I try not to think too hard about it, focusing on the first impressions that bubble to the surface, and scribble down a bulleted list of responses:

- How is climate change shaping my reproductive life? It's not. What shapes my reproductive life are: ambivalence; being single; my age; social and family networks; concerns about racism, politics, the state of our society, and now Covid. I am more concerned with Covid than anything, since it seems like it will never end. How would I deal with this if I were pregnant? If I already had a kid?

- Climate change. Reproductive life. Myself. Hmm. I talk about climate change and reproduction all the time. But for myself? Hmm. I guess if I have to think in personal terms, it's like a confirmation that I'm making the right choices. That not having kids, at least not my own biological kids, is a very good thing for me. My anxiety is already high with this wildfire season; it would be through the roof if I had a kid to worry about . . .

- But then again, if I had a kid or kids, would I be as fixated on wildfires? Wouldn't I have other concerns, like keeping them from eating too much sugar, or wondering how they grow out of their clothes so fast, or figuring out how to get them socialized with other kids in the middle of a pandemic? Would they be the only Black kids in their class? Would they have enough cultural role models? Would I be worried about how little they would see their cousins, who live too far away? Would I think about climate change at all if I had a little baby to occupy my thoughts? Or would I think about it so much that I would be consumed with guilt?

- What the hell would I have to feel guilty about? This is the oil companies' fault. This is our ineffectual representatives who refuse to see this as a crisis. They want me to feel guilty, to feel responsible, to feel like climate change is my responsibility. Bullshit. This rests squarely with them. But what does that have to do with kids?

- Haven't people with much bigger challenges still formed families and been happy? Those families probably made them feel safe from all of the societal stuff. Then again, how many families reject their children and make them feel isolated? Family isn't always the shelter where you weather the storms. Sometimes it is the storm.

- Why do I even have to deal with these questions? As a woman, I think we've gotten to the point where we should no longer have to assume that having a baby is a requirement. I can be worried, anxious, stressed, scared, and everything else with respect to climate change. That's valid on its own. Why does it all have to come down to what is or isn't happening in my womb?

While writing, I'm reminded of a summer night in my childhood, when I was maybe eight or nine years old, visiting my grandparents for the weekend. My sister and I loved going to their house in the San Fernando Valley, nestled amid the craggy, rocky mountains and dry desert landscape, because it marked a distinct contrast to the cramped Inglewood apartment where we lived with our parents. That particular night, I fell asleep to the sound of strong winds whipping around the boulders of the nearest peak. I woke up a few hours later feeling strangely warm for no apparent reason. The heat wasn't on—it was the dead of summer. Even stranger, it sounded like it was raining hard outside, which I knew couldn't be true, but in my blissful ignorance I chalked it up to a sudden hot thunderstorm and went back to sleep.

The next morning when I looked out the window, I was shocked to see that the entire mountainside next to the house was blackened, charred entirely from tip to toe, and that the burn marks ended just a few meters short of the homes in my grandparents' little subdivision. So that was what the noise had been—not a freak rainstorm in the middle of the night but the opposite: an enormous fire, big enough to swallow the whole mountainside and to heat the house from outside like a furnace. I leapt out of bed and raced downstairs to the kitchen to inform my grandmother that there had been a fire. Unruffled, she looked up from her newspaper and said, "I know. We were out there fighting it all night with the neighbors. And then the fire department arrived and put it out."

Remembering that incident this day, I was stunned. My grandparents were out in the middle of the night fighting a wildfire? What would I have done in that instance, if I'd had a child of my own? Of course, I like to think that I would have banded together with the neighbors, calmly going into problem-solving mode while (I hoped) my child slept peacefully. In reality, it might not have gone so well. I might have panicked, throwing my kid and all that I could carry into the car, racing off into the night and away from the danger. These are the questions I can't answer: Would parenthood make me desperately selfish? Would I retreat into a world of me and mine against the elements, leaving others to their own devices? Climate-driven kid questions demand reflecting on what our responsibilities are to each other, not just to our own children and families if we have them but also to our communities and the broader society. What do we owe each other? Can we keep each other safe in the midst of this kind of crisis? Given how fractured our nation is politically and socially, I'm not sure that we will. But then again, I can't be sure. None of us can.

The house party reconvenes, and we are organized into small groups to share what we've written. By now I'm deep in my feelings, less the detached researcher than the still-wary person who doesn't know how open to be with the rest of the group. We get into our small groups and I'm paired with just one person, who is also Black, and our conversation is so sweet, and sad, and yet really comforting. Our feelings find a mirror in each other, over and over again. I feel really seen. All of my worries and feelings, trepidations and yearnings, my hopes for the future and for Black children and all children—all of it is held and reflected back to me with such understanding that it brings me to tears. I am not alone. I realize now that it is the retreat into silence, the not talking about it and the assumption that we can think and argue and analyze and dismiss our way out of it—that is the problem. No one is here to solve the problem for me, but maybe my feelings aren't problems to be solved as long as I can acknowledge and express them out loud. When I do that, I find understanding and recognition are waiting for me.

We leave our pair and rejoin the larger group. As we start to talk, I am shocked to find that the deep, intractable ambivalence I have long felt about having children is the common feeling here. No one has made up their mind, everyone is uncertain, we all feel that these questions feel overwhelming, given the circumstances. Some people in the group can't imagine having children grow up in a society that has undertaken such a hard political shift to the right. We all worry about Covid-19, whether it is here to stay, whether worse is yet to come. Climate change is the container for all of it, the framework of our fears about a future that we all, in our own way, perceive with no small sense of horror. My feelings are specific, but they are not individual. They are shared by a group of people who on the surface seem really different from me. And they are shared far beyond our little group of ten. As the next chapter demonstrates, these feelings are shared by thousands of reproductive age people around the world. Maybe more than thousands.

At the end of the day, this is my takeaway: emotions that seem internal and individual have to be brought out into the public, or at least into a group beyond yourself, to make the important connections to social and political systems beyond yourself. This is a necessary first step to understanding that although feelings aren't problems to be solved, they can be the basis for something powerful, whether that power is to be found in an emotional connection to a stranger or in political actions, or both. This space that we've found at the house party is powerful too; it has made room for us to share our feelings, find recognition from others, and feel seen, heard, and understood. I'm still deeply anxious about climate change. It helps to know that I am not alone.

6   Reproductive Resistance
in Public

In late 2018 the Intergovernmental Panel on Climate Change (IPCC) released a special report.[1] The report detailed what the impacts of global warming would be at 1.5 degrees Celsius, with a comparison of what it would look like if that target wasn't met and instead we reached 2 degrees Celsius warming. The report's ninety-one authors cited more than six thousand scientific references. It was written in the usual dry, measured tones of climate science reports. And it set off a firestorm among activists.

The document said that we've already reached 1.0 degrees warming above preindustrial levels and that we're feeling the impacts in extreme weather, heat waves, sea level rise, and melting sea ice. Those impacts are not going to go away—they're baked in and will continue over the coming centuries or millennia. But the report also said that those effects would get worse—way worse—if warming exceeds the 1.5 degree target. Threats to human health, livelihood, food security, and water supply, not to mention sea level rise, heat waves, drought, and species loss and extinction, will all intensify in the coming years, *no matter what.* But if we exceed 1.5 degrees and get closer to 2.0, some of these changes will be catastrophic, possibly irreversible. And, going at the current rate, we were on track to reach 1.5 degrees pretty quickly—potentially as soon as 2030. At the time, that was twelve years away. The document concluded that limiting global warming to the 1.5 degree target would require *"rapid and far-reaching transitions* in energy, land, urban and infrastructure . . . [transitions that] are *unprecedented* in terms of scale . . . and imply deep emissions reductions in all sectors."[2] And, as if that weren't enough, human actions would have to change at a level that would cause carbon emissions to fall by 45 percent from 2010 levels by 2030, and reach net zero by 2050.

Unprecedented. Rapid and far reaching. *Twelve years.* The urgency of the report leapt off the page. And yet, while the report was saying that the time to make these policy changes was now—as in *right now,* like *right this minute*—Donald Trump, a notorious climate denier, was in the White House. He had pulled the United States out of the historic Paris Agreement, the first to set binding targets toward achieving shared global goals to limit global warming. Because of Trump's actions, the US government was no longer legally bound to meet any of the targets of the agreement. And if the United States wouldn't meet those targets—given its status as one of the biggest carbon emitters on the planet—the rest of the world could not meet them either.

DON'T GET COMFORTABLE

Several months after the IPCC report was released, Alexandria Ocasio-Cortez (aka AOC) was interviewed at a Martin Luther King Day event, where she spoke about young people's frustration about climate change. They were fed up, she noted, with older adults' and politicians' slow responses to creating bold, innovative policies for mitigation and adaptation, and their preoccupation with the financial bottom line. She said: "Millennials and people, you know, Gen Z and all these folks that will come after us are looking up and we're like: 'The world is gonna end in twelve years if we don't address climate change and your biggest issue is how are we gonna pay for it?"[3]

The twelve years she was referring to were those outlined in the IPCC report. Government leaders, she argued, were dragging their feet on the response, to the point of total irresponsibility. The report was shocking—damning, even. Not only did it address the scientific context of rapid ecological change, it clearly pointed to an immediate need for political change. The kinds of policies that would shift economies away from reliance on fossil fuel development and toward alternative energy, well-paying sustainable jobs, and the like. However, politicians weren't taking those actions on either side of the aisle, instead choosing to be preoccupied with mundane budgetary concerns while the planet was poised for catastrophe. "Right now," AOC said, "with the current administration, with the current circumstances, with the abdication of responsibility that we've seen from so many powerful people, even people who abdicate that responsibility by calling themselves liberal or a Democrat, or whatever it is, I feel a need for all of us to breathe fire." In other words, the time for progressive climate action from politicians was overdue. And for young people, the time was ripe for rage.

Less than a month later, AOC stood in her kitchen, cooking dinner and talking to her three million followers on Instagram Live about the climate crisis. She argued that children's lives would be very difficult in the future and that, as a result, young people were asking themselves from a moral perspective whether it was still okay to have children. Arguing that the lack of urgency among her fellow politicians would kill us all, AOC noted that for young people concerns about having children were no longer just issues of financial or career-stage readiness: they were about climate anxiety. Her words were jarring. What she raised publicly—and a lot of young people struggle with privately—is a rethinking of what for many is both an important rite of passage and a symbol of hope for the future: raising kids and leaving a human legacy. This is inherently a political issue, underscored by the fact that AOC, a politician, was clear that we wouldn't have to ask these questions if elected officials had taken the necessary, aggressive policy steps to address climate change.

Of course, AOC was not the first young woman to publicly link parenting concerns to the climate crisis, nor would she be the last. Within months of her Instagram Live message, two public campaigns appeared online: #NoFutureNoChildren and BirthStrike. Both campaigns were led by young women: one by a Canadian Gen Zer, and the other by two British Millennials. Both campaigns argued that parenting, reproduction, and climate action are closely tied together for young people who are worried about the state of the planet. In both instances, the campaign creators were climate activists who had read the IPCC report and were struck by the seeming-finiteness of a twelve-year deadline for averting ecological catastrophe. And both campaigns arose from the immediate question the deadline evoked: What about having children? Like the leaders of Conceivable Future, these young women were engaging in reproductive resistance, as outlined in chapter 1. However, they were not asking whether people could—or should—become parents. They weren't asking questions at all. Instead, they were firmly declaring: *we will not have children until our government leaders take effective action to save this planet from the ravages of climate change.* Seeing their declarations online, thousands of other young people agreed with them and declared the same.

Both campaigns closed within a year or two of being launched, in large part due to controversy and resistance from media, ideological opponents, and, ironically, their own supporters. This chapter traces their trajectories to understand what happened and why reproductive resistance on environmental grounds continues to generate so much controversy. Unlike public campaigns that linked reproductive concerns and environmental issues in past decades, both #NoFutureNoChildren and BirthStrike were either created or co-led by women of color, and their leaders were explicitly antiracist in their approach. Both campaigns sought to shift the focus of the conversation to a critique of government systems and policies. They argued that rather than babies and children being dangerous for the planet, climate change is making the earth an increasingly inhospitable place for human life. They specifically rejected arguments stating that "overpopulation" in general—and the reproduction of people of color, poor communities, and those in the Global South in particular—is driving climate change.

So why did these campaigns shut down? Despite their best efforts, BirthStrike leaders lost control of their own messaging. Their narrative was taken over by those who believed that the only way to talk about climate change and children is to tell people not to have them. Although #NoFutureNoChildren did not end in exactly the same way, its leaders also had to contend with misunderstandings, accusations of population control, and "supporters" whose goals veered toward the hateful. The leaders of both campaigns underestimated the vigor of antinatalists on both the left and the right, as well as how often other progressives would misinterpret their approaches to reproductive resistance. They soon found out. And it was this relentless misinterpretation that would prove to be their undoing.

NO GRANDCHILDREN FOR YOU

Imagine you're eighteen years old. You're starting your first year of university, but it's hard to concentrate on this new experience. Your mind returns to the IPCC scientific report, the one warning that we only have twelve years to keep temperatures below a crucial climate threshold, and you can't ignore your feelings of hopelessness. How can you to go to school, make friends, and think about a future where you might find a job, a partner, and settle down, when the future of the planet is so uncertain? Even more frustrating is that your parents and their generation don't seem to see the urgency. It's not that they haven't heard you talk about your climate emotions, specifically distress, sadness, and despair; you make your feelings heard to your family, teachers, and community members, loudly and clearly. But they just don't seem to take you seriously. They applaud you for your activism but don't engage in any of their own. They elect the same leaders without holding their feet to the fire on climate policies. In fact, by and large, they don't even know what elected officials' climate policy proposals are. And when it comes to your anxiety and grief about the planet, they simply tell you that your generation is going to get us all out of this mess.

You have your whole life ahead of you, and it doesn't sound like a great one: devastating fires, scary storms, floods, rising temperatures, and government leaders seemingly more entrenched in business as usual. Even worse, this has already been your whole life. You were born into a world that was already being turned upside down by greenhouse gas emissions and their concentrations in the atmosphere. And older generations have been neglecting, ignoring, dismissing, downplaying, and denying it the whole time. What do you do? How can you get older people to stand up and say enough is enough? To make them feel your overwhelming sense of worry that maybe everything will not be okay, and that—given the circumstances—planning a future with a family in it makes no sense at all? What do you do? You resist.

Emma Lim's activist trajectory encapsulates exactly this kind of resistance. At eighteen years old, she was an Asian-Canadian activist who could mobilize hundreds of thousands of her peers in national demonstrations; in fact, she had already done so while she was in high school. She moved into direct action around 2015, when she first participated in student walkouts over changes to Canada's national sex education curriculum. Later, as Emma became more interested in climate change activism, she was inspired by Greta Thunberg's school strike, which led her to wage a solo protest every week as part of Thunberg's Fridays for Future movement. Over time, Lim joined up with friends, two or three of them striking each week, and later she staged larger climate actions in Toronto and Ottawa.[4] Eventually they founded Climate Strike Canada, and in September 2019 they organized Canada's national School Strike for Climate, an international event that attracted between 315,000 and 500,000 participants in Montreal alone.[5]

A couple of weeks before that, Lim and her friends had launched another, smaller campaign: an online pledge called #NoFutureNoChildren, with Lim as the public face. Their message was deceptively simple: "I pledge to not have children until I am sure my government will ensure a safe future for them." However, the impetus behind the pledge was emotionally and politically more complicated. Writing about it on her website, Lim explained: "I am giving up my chance of having a family because I will only have children if I know I can keep them safe. It breaks my heart, but I created this pledge because I know I am not alone . . . we've read the science, and now we're pleading with our government."[6]

Prior to launching the pledge, Lim had spent hours poring over the IPCC report, feeling a mounting sense of fear and sadness that would not go away. She had turned to her friends, their group chat sowing the initial seeds for their activism.[7] Their feelings about climate change—what Emma described in an interview as "deep eco-anxiety"—and the ways adults often dismissed them were the real impetus for the campaign. They wanted to capture older generations' attention, and they knew that sharing scientific facts and figures wouldn't work; they had already done that both privately and publicly, and it hadn't swayed those closest to them, not to mention strangers. What it came down to was the fact that their emotions were involved, and they wanted older people's emotions involved too. How could they achieve that emotional pull? What would make older people realize that climate change was an immediate threat that was upending young people's lives?

Throughout their conversations, there was one thing that emerged, the thing their parents, teachers, and other older adults cared about, talked about, and regularly pressured younger family members about: children. Older people want grandchildren, they concluded. The thought of not having them makes them panic, or at least it disrupts some of their taken-for-granted assumptions about the kind of families they'll have in the future. These are exactly the kinds of ways climate change disrupts expected future families for young people who struggle with climate anxiety and the kid question. So Lim and her friends created their pledge, intending to spark a cross-generational dialogue about how climate change is stealing the future families many young people want. Leveraging the media savvy and organizing tools they had developed through their national climate organizing, they launched the pledge just before a national election.

Although the pledge itself focuses on government action, their ultimate goal was not to reach political leaders; instead, it was about sparking a conversation. They wanted young people to see the pledge and sign onto it. From there, they hoped pledgers would share it through social media so that their family members, their parents and grandparents in particular, would see it and consider the effects of climate change as they went to the polls. "At least in Canada, the majority of people don't really consider issues until they're impacting them," Emma argues. "And if you're pretty well off and live in a nice community, you're not really going to care about climate change unless maybe it's your own child saying you're not going to have grandkids because of climate change. And then all of a sudden you might start to read articles that you previously would have scrolled past. So it's that kind of thing, this pledge, we hoped it would help young people's eco-anxiety get a wider platform, and also we wanted to really impact the older generations who are really concerned about young people having kids. And then, political action comes mostly as a result of what the people want when a certain number of people want it."

Vice picked up the pledge immediately and posted it to their international release page, prompting a number of other outlets to pick it up too. From there, the responses started rolling in. Lim and her friends were shocked: they had guessed that it would attract around five hundred signatures and remain local to Canada. Instead, people responded in droves, including from abroad. Within a week there were more than a thousand signatures and comments on the pledge. Nine months later, when they closed the website down, the pledge had ten thousand signatures from countries all over the world. #NoFutureNoChildren leaders were surprised, not just by the international response, but by the range of ages and the intensity of some young people's climate emotions. "We analyzed it just to see what the comments were, and we were like, wow, this person's really young, this person's really old, this person is clearly just trolling us," Lim told me. They were particularly struck by the responses from people as young as thirteen years old, because their responses were the most nihilistic: "People in their thirties were like, 'Yeah, it's really rough, I really love kids, but I'm considering not having any.' And then the thirteen-year-olds were like, 'Yeah, I wish I wasn't born. I'm anxious all the time, and I don't want kids.'" Responses to the pledge are currently housed offline in an Excel spreadsheet on Emma's computer. She scrolls through it as we chat, occasionally reading them out loud:

> So one said, "I took this stance at twenty-two because the crisis I saw facing the earth."
> Oh, this person's thirteen. They wrote, "I'm taking this pledge because I won't leave my children dying in a problem my parent's generation could have stopped."
> This person said, "How can I bring a child into a world that is breaking down?"

The experience of climate change and the emotions it provokes are deeply tied to social, political, and economic conditions—the same conditions in which people decide to become pregnant, give birth, and raise children. While it may seem that Emma and many of those who signed the pledge are opting out of having children, they're actually stating that they feel they are left with no choice about having them. Lim, for example, has always wanted children; she even worked as a nanny in the past. But the climate crisis has changed the conditions of life in the present and the future. Building on the personal-as-political perspective found in Conceivable Future's work, Emma's public advocacy is predicated on refusing to keep distressing climate emotions, and their impacts on reproductive questions, private. The emotions arise from shared climate impacts as a result of living with national economies based on fossil fuel extraction—and in the #NoFutureNoChildren campaign, those impacts are actively and openly resisted, to place the responsibility back on the shoulders of elected officials and the older generations who keep them in office.

GOING ON STRIKE

BirthStrike was a UK-based social media campaign that operated publicly from March 2019 until it closed in August 2020. It was initially created in late 2018 by Blythe Pepino, a white British Millennial, musician, and climate activist, after she read the IPCC report. Like #NoFutureNoChildren, BirthStrike focused on resisting having children based on the state of climate change and ineffective government action to address it. However, this campaign was more interactive; it began with a Tumblr page and expanded to other social media platforms, centering primarily on community discussions in a Facebook group. The format allowed for more conversation with other like-minded young people who were anxious about climate change, nervous about having children, and needed others in a similar position to talk to about it. In fact, this was the goal of the group: to create space for people in their reproductive years, who were fearful of bringing children into a moment of climate collapse, to come together in solidarity. Unfortunately this level of engagement and dialogue was one of the things that derailed the organization's efforts.

From early on, BirthStrike had a clear political mission. After reading the IPCC report, Blythe was devastated by its dire forecast. This led her to attend her first meeting at a local chapter of Extinction Rebellion (or XR), a UK-based environmental activist organization. XR views climate change as crisis for life on earth; its webpage describes the stakes of climate change as "social and ecological collapse" and "mass extinction."[8] In other words, they frame it as an existential crisis, hence their name. This proved to be important for how Blythe would develop the approach and mission of BirthStrike. Her immediate response to both the report and the meeting were visceral: "I came to this decision—that I couldn't face the idea of having a kid," she noted. "Then I got quite active talking to a lot of people who I knew understood the situation with the ecological collapse. I soon realized that a lot of people were feeling the same way, both men and women."[9]

Blythe partnered with Alice Brown, another white Millennial, to start BirthStrike as a media-focused movement of Millennials and younger youth. Their official statement read: "The threat of ecological and civilization collapse is certain unless we transform our systems as a global community, justly and swiftly. Our future and that of our children is in the balance. Those who have signed up to #BirthStrike are raising awareness by saying that this is now affecting the human ability and desire to give birth."[10] Blythe's and Alice's goal was to bring attention to the urgency of climate crisis by highlighting their fears of raising children in a society rooted in systems that are leading to ecological collapse. They wanted to help shift the general conversation about climate change from a focus on the future to one situating it as a present-day existential crisis—a crisis that has long-term implications for their lives, their families, and their legacies on the planet. One way of doing it was to highlight the emotional devastation caused by climate change. At the same time, the core of their argument was economic: something is deeply wrong when a society prioritizes economic growth over the safety of its citizens, particularly its children.[11]

After launching, BirthStrike followers quickly grew to number seven hundred, operating across various social media platforms including Facebook, Twitter, Instagram, and Tumblr. Followers really resonated with how the campaign connected eco-emotions to climate change and the kid question; for some, the emotional turmoil was the most important part. One follower described her feelings thus: "For me, the biggest factor is the stress and depression around our removal from a connection to the planet that we live on, the natural world. When our kids would get to, say, thirty, what would that world look like?" Another young activist described the moment when the issues struck home for her: "The past New Year's Eve, a group of us were playing a game where we just asked questions that people answered. One of the questions was, 'What are you looking forward to in the future?' I surprised myself when I said, 'I want to have a future where I even have a choice whether to have a child or not.' I burst into tears, and everyone else burst into tears, because we realized that it had got to that point."[12]

Modeling their message after that of XR, BirthStrike framed climate change as an immediate existential threat. They developed a distinct platform with a detailed, progressive mission statement that rejected population control, instead prioritizing equality and standing in solidarity with parents and would-be parents. Their mission was as follows:

> We are of all identities from around the globe. We cite many different reasons as to why we have each signed up but we all agree on the following:
>
> • Life, as we know it in the Holocene, is in great, immediate danger.
> • We stand in solidarity with the environmental justice movement, the academic and scientific community who demand fast acting and transformative system change towards an equality based, sustainable, caring and non-violent future for humans and all life on earth.

- BirthStrike stands in compassionate solidarity with all parents, celebrates their choice and fights for the safety and lives of their children.

- BirthStrike disagrees with any population control measures and recognises the colonial violence of such measures having been proposed in the past and present.

- We disagree with focussing on the topic of population before equality based system change in regards to tackling the environmental crisis.

- By adult we mean the typical age of attaining legal adulthood which is 18 years old.

- BirthStrike is a voluntary group and anyone can join or leave at anytime.

- BirthStrike is inclusive of all capacities to conceive, parents and those who are childfree, all genders and identities.

From the start, their stance provoked endless discussion across social media platforms, both from those who aligned with BirthStrike's perspective and those who insisted that population control is in fact a necessary environmental strategy. After several months of these discussions, and after educating herself further on the racial dynamics of population and environment debates, Pepino put out a statement meant to tackle discrimination head on and quell any further question about the movement's values. She wrote: "A big factor in why BirthStrike has received so much coverage is because . . . I believe, the fact we started out looking mostly like 'white women from the UK.' However, we are now a global multicultural community and we should show that we are proud about that . . . so that anyone and everyone can feel welcome to join the conversation, find solidarity and share their story to call for #systemchange."

The statement invited people of color and those from countries in the Global South to lead the conversation, to help ensure that the organization's messaging was inclusive and could accomplish what

they intended. It also unequivocally declared BirthStrike's nonjudg-
ment of pregnancy and birth, and their goal of drawing a boundary
around their messaging, specifically by rejecting ecofascism. "We
need to prevent the movement being used to create a 'hostile environment' and fresh borders around birth," they wrote. "At worst,
already oppressed women could be pressurised or forced into not
having children 'for the environment' so as to regulate population for
fascist ends . . . these are some of the serious reasons we do not campaign with language that seeks to coerce, and we do not campaign
for population control as a fix for the environment, and why we
remain a non-judgmental movement."[13]

Despite BirthStrike's core message resonating with many of its
followers on a personal level, some supporters were primarily concerned with the public stance on population. Specifically, they
wanted BirthStrike to advocate against more babies on earth. It
didn't help that when Blythe and Alice showed up for media interviews, it turned out that journalists already assumed that they were
coming from an antinatalist approach. Population organizations
reached out, asking to partner with them, while the articles and television appearances they participated in—including one with Tucker
Carlson on Fox News—felt hostile, like willful misinterpretations of
their message. Perhaps their actual message about systems change
was too complex for the superficial ways media outlets wanted to
engage with them; perhaps neo-Malthusianism is just too deeply
embedded in mainstream environmentalist thinking and values.
Perhaps some people were turned off by two middle-class white
women talking about reproduction and climate change, and assumed
that the only way they could do so was from a perspective of privilege. Blythe actually welcomed this last assumption—she appreci-
ated being able to get her views published in mainstream outlets like
Cosmopolitan, where she could surprise people with an unexpected,
progressive message.[14]

Regardless, BirthStrike became deeply mired in the controversy,
mainly in the form of tense debates across their social media

platforms. Brown left her leadership role and was replaced by climate justice activist and writer Jessica Gaitán Johannesson, a queer woman of Colombian, Ecuadorian, and Swedish extraction. Together Jess and Blythe worked to hone BirthStrike's message to focus on climate justice, with an explicitly antiracist framing. However, it was too late. Members of the Facebook group doubled down on their efforts, posting hostile, antinatalist articles and declaring, over and over again, that the group needed to declare an unequivocal stance against population growth. These efforts were nothing new within the environmental movement—antinatalists had pressured the Sierra Club to do the same in the mid-1990s and again a decade later.[15] For BirthStrike leaders, however, the relentless pressure created a chilling effect that made them hesitate to create new posts and discussion threads. Given that their goal was to have ongoing dialogues about reproductive anxiety and to root this perspective in calls for changes to economic systems, ceasing the dialogue was counter to the entire point of the campaign.

They were up against an organized environmental movement of ecofascism, powered by white supremacy and racial hatred. Ecofascists cite population growth, environmental degradation, and the rapidly growing climate crisis as reasons to target immigrant communities of color. Feminist scholar Betsy Hartmann refers to this as the greening of hate—the ongoing scapegoating of racialized immigrants for environmental degradation.[16] As writer Sarah Manavis writes, ecofascists thrive in online spaces: they are loosely connected through virtual communities where they embrace such disparate values as "veganism, anti-multiculturalism, white nationalism, anti-single-use plastic, anti-Semitism, and, almost always, a passionate interest in Norse mythology."[17] And because they are already in those spaces, it was easy for ecofascists to discover BirthStrike and infiltrate it with their extremist messages.

To be clear, it's not just that ecofascists are antinatalists. Rather, their perspective is neo-Malthusian at the farthest extreme: they believe that the size and growth of human population strains natural resources and that large numbers of displaced people, whether immigrants or refugees, are a threat to social, political, cultural, and most importantly, environmental stability—and they believe that the scale and urgency of the climate crisis justifies a violent, even murderous, response.[18] Although it may appear to be a new movement, ecofascism has its roots in Nazi Germany, where values rooted in deep ecology, conservationism, and white supremacy aligned.[19]

BirthStrike's struggles with ecofascists and other antinatalists made it clear that, despite Blythe's and Jess's best efforts, there was a disconnect between their goals, messaging, and strategy. Increasingly, their intentions were obscured as media and members of the public challenged Blythe's and Jess's credibility, sometimes using climate denial to reject their arguments; this "epistemic injustice" ultimately led them to close BirthStrike in August 2020.[20] In their closing statement to supporters, the leaders wrote: "Having spent the last 6 months in a process of reflection, the BirthStrike team have come to realise that BirthStrike is no longer tenable as a campaign tool for action on the climate crisis. In the light of the recent BLM protests spurring conversations around racial and social justice, and what we see as a welcome and increasing awareness of the inextricable links between social/racial justice and climate collapse, we think it will be dangerous for us to continue to campaign as 'BirthStrike' going forward."[21] Their reasons for discontinuing were as follows: first, the name BirthStrike received a lot of harmful attention that ultimately served as a distraction from their goals; second, they worried that centering the emotions of fear and anxiety might feed into overpopulation narratives and ecofascism, exactly what they did not want to happen; and third, they had come to realize that focusing on personal reckoning with climate collapse and climate grief runs the risk of focusing on the future in ways that blind us to the real suffering many people, particularly low-income communities and those in the Global South, are enduring right now.

That wasn't the end of their advocacy work, however; it was simply time to move on and grow in another direction. In September

2020, Blythe and Jess created a private support group called Grieving Parenthood in the Climate Crisis: Channeling Loss into Climate Justice. Their focus on grief and loss highlighted the core eco-emotions at the heart of reproductive anxiety while heading off debates about population. They made sure to notify their followers that "the new group is for those who agree justice must be front and centre of any action taken. We hope with this new focus the new group will be more collaborative and aligned."[22] The statement proposed a more mindful approach to supporting followers in the new group, where the focus could adhere more closely to the key issue at hand: sharing space to process young people's deep, collective climate grief as they rethink their dreams of parenting as well as advocating for a justice-based approach to collective climate action.

RESISTING THE WRONG NARRATIVE

Before launching #NoFutureNoChildren, Emma Lim and co-organizers were concerned about how their pledge might be met with controversy—specifically racial controversy. Emma has co-organized with peers of all racial backgrounds, and she and her cohort are sensitive to the history of population control in Canada. So they determined that they would address it in how they crafted their pledge from the start. They worked with sensitivity readers to ensure that the language would not come across as promoting population control, which Emma decries as a "white supremist myth." They recognized that particularly for minoritized racial and ethnic groups, having children, passing on traditions and cultural legacy are very important. "That's just as powerful, or even a more powerful way of fighting climate change and trying to enact systemic change," Lim says. "We knew people of color would care about this. That came up a lot when we were creating the pledge. Like in Canada, there's been forced sterilization of Indigenous women, and so we were like, how can we make sure that this doesn't come across as, 'You shouldn't have children'? Like, because that's horrible." Although the pledge did not collect information about racial or ethnic identity, Lim assumes that most of the people who signed it were likely white people in their twenties and thirties. "It's a really complicated issue, and I think it's not a bad thing that this pledge is more popular among certain groups compared to others," she argues. "Because for those others, having kids is really important. Reducing the number of people on the planet won't work. We need systemic change, from the bottom up and from the top down. We need new systems."

The #NoFutureNoChildren pledge did not include a statement about population, but Lim has strong feelings about the subject. "People who say that population is the problem, my opinion, I've seen a lot of antinatalism come my way and I don't agree with it," she says. "At the core, the issue is not that we don't have enough resources for all of the people. The issue is that we're hoarding resources. We throw out so much food, we consume so much in general. So yeah. If you think antinatalism is the solution to the problem, I feel like that's so short-sighted, it's ridiculous." BirthStrike leaders mentioned population control in their closing statement on the end of the campaign, noting that they had initially thought that they could play an important role in debating and resisting overpopulation narratives, but quickly realized the campaign's name inadvertently contributed to harmful narratives. Even pointed, critical messages rejecting neo-Malthusianism can be, and often are, overtaken by those with a population control agenda.[23] Unfortunately, as in the case of BirthStrike, this removes progressive perspectives from the discussion and cedes the conversation to the loudest and most dangerous voices in the room.

WHAT NOW?

Against all odds, Emma, Blythe, and Jess still have a sense of hope in the future, including the possibility of someday being parents. When

I ask Emma how she feels about having children someday, her response is "tentatively optimistic." It makes sense; some level of optimism, or hopefulness at least, is probably necessary to fuel her ongoing climate activism. Of course, she has also stepped away from #NoFutureNoChildren, which has allowed her some breathing room to focus on things like attending university, raising a cat, and building a relationship with her girlfriend. She is hopeful that necessary systemic changes will be enacted for the climate, and is committed to working toward those changes, but not through a campaign focused on reproductive resistance. The same is true for the leaders of BirthStrike. After closing BirthStrike, Blythe tells me in an interview, she felt drained by activism but had reached a certain point in her journey of climate grief where she started to feel "a level of delusion . . . in order to maintain psychological health." That led her and her partner to consider having a child, though unfortunately a health challenge upended that plan. "I had come to a Zen position of 'life is suffering' and an acceptance of the kind of suffering climate breakdown could bring to that child," she says. "So I would like to have a child, and if that health issue wasn't happening I would probably be planning to have a child, I think. It still feels like cognitive dissonance from day to day."

Jess and her partner also eventually began to explore the possibility of having a child, even as they struggled with eco-anxiety. "It's horrible some days," Jess says, "because the fear doesn't go away. I thought about it, like, why do people keep having kids in the midst of existential crisis? And my partner said, 'By not thinking about it.' You have to be realistic and then emotionally you have to step away sometimes." Some months after this conversation, Jess published a book of climate essays and an article that made it clear that she was in fact still thinking about it, very much so. "It's not that I feel more optimistic," she wrote. "Rather, I've come to think that in trying to decide whether it was right or wrong to bring a child into this now, I was asking the less helpful questions, the ones less anchored in the now I'm living in . . . the way forward must never lie in giving birth, but to create all kinds of families, and for those families to look after each other."[24]

If followers of their campaigns are surprised to find that Emma, Blythe, and Jess would like to have children, they've misunderstood them all along. Their campaigns, rendered in the language of public reproductive resistance, were never about a decision to not have children. They weren't final declarations. These were conversations being had in public with thousands of other people. There was no declaration of never-having-kids, mainly because they hadn't made those decisions themselves. Instead, they were sounding the alarm on what this climate crisis is doing to young people: plunging them into a state of deep uncertainty and fear, disrupting their ability to imagine a hopeful future—including future children—because of their horror and grief over where we already are in the present.

Public reproductive resistance focused on climate change is messy, driven by emotions and politics. This is the entire point: it stems from climate anxiety and reproductive anxiety, as well as a refusal to deal with these challenges privately when they stem from public, political decisions. But it is also a tense, fraught topic, because reproduction itself is always tense and fraught in the public sphere. Competing social values, laws, and the history of how socially privileged groups have shaped the contours of the public debate all come into play with how these ideas are voiced in public. Reproductive resistance still sits at the nexus of difficult debates around reproductive autonomy, the significance of human numbers, and how those numbers impact the earth's ability to sustain life. But now, it adds complicated emotions to the debate, which have likely been there, under the surface, all along.

Conclusion

NO CLIMATE JUSTICE WITHOUT
CLIMATE EMOTIONS

The emotion I haven't written much about in this book is love, although it lurks at the edges throughout. My fierce love of a clear blue sky on a sunny day, the craggy mountains in the distance and boulder-covered hills nearby, bees at the lavender bush while hummingbirds flutter around the weeping bottlebrush tree out front—this is a love borne of its opposite, the knowledge that these things may not stay.

When I first wrote these words, we were in the midst of a heat wave that brought temperatures between 105 and 108 degrees Fahrenheit, with daily rolling blackouts. My city had begun enforcing water rationing; all of the plants in the backyard were dying. They came back the next spring though, this time drowned in a deluge created by a series of atmospheric river storms that pummeled the West Coast, bringing cold rains, floods, and a very disorienting sense of wondering what it was all coming to (particularly the day it snowed in my city). But even the winter storms can't change the fact that summers in Southern California have become progressively hotter, drier, and more fire-prone over the past decade or so, a trend that shows no sign of reversing.

I do love this dry Southern California landscape though, despite its challenges. We have some of the worst air quality in the nation, which is concentrated in Black and Brown communities in the region. Diesel-powered trucks traverse our freeways as warehouses and logistics centers replace what were previously citrus fields. A lot of the children here have asthma. Adults have chronic cardiovascular and respiratory conditions. Environmental justice organizations have been working for decades to reduce community members' exposures to polluted air. Having joined the fight with several community-based organizations in the region, the task feels monumental, at times overwhelming.

Love and grief. Determination and outrage. They are all part of climate anxiety for me, but I keep coming back to love. Connecting to the things we love is important while we deal with the climate crisis. Whether it's children, a sense of place, community, or our own creativity, love can help us navigate and even transform the emotions that might otherwise make us feel powerless. The other day, I was talking to Richard, a twenty-year-old college student from New Jersey who is majoring in filmmaking. To my surprise, he said that he felt that climate change made his degree a waste of time, despite how much he loves it. He noted: "I had sort of a crisis of thinking about, like, why am I majoring in film when the world is ending? And like . . . that will most likely have really very little purpose for the future. I don't know what it would have use for . . . it's kind of scary to think that what I'm currently working towards may not even be available in the future, or may not be important in the future. Because the way I'm thinking, I'm just assuming that we're not going to have lives that are able to sustain luxury and entertainment. Who knows, we may not have time to even watch movies in the future. So why am I doing this? How am I contributing anything?"

I'd heard this kind of sentiment before, from a twenty-two-year-old recent university graduate, a woman who had studied music and spent most of her time writing and performing songs with her band. She felt guilty about dedicating her energy to the art she loves, when she felt that a major in environmental studies might have been more useful. "Maybe I should be becoming an engineer," she said, "or a conservation studies professor or something. Like, how frivolous is it to do music? How is that going to help the climate crisis?" But she was wrong, as was Richard. Art is exactly what we need right now, and so much more of it.

We need art like *The North Pole*, an online political comedy television show created by Movement Generation, a climate justice organization led by young people of color. The show skewers gentrifiers and elitist food movements, pokes fun at the drivers of climate change, and uses satire and political commentary to spark a conversation about climate justice in low-income communities. We need mainstream films like *Don't Look Up*, the 2021 satire ostensibly about a comet heading toward Earth (but in fact the film was really about catastrophic climate change); its hilarious sendup of how politicians and media personalities consistently ignore or obstruct efforts to address climate-related emergencies was spot on. We need movies like *First Reformed*, a somber film about a young priest who tries to help a small community cope in the aftermath of a tragedy borne of extreme climate distress.

We need documentaries to help us learn how to live in relationship with the Earth, like *Remothering the Land*, a film about Indigenous regenerative farming practices that support healthy soil, people, and animals. We need documentaries like *Katrina Babies*, which highlights the long-term effects of storms like Hurricane Katrina among the communities most impacted by them. We need films about reproductive justice too, like *Aftershock*, a documentary about Black women and maternal death in the United States, as well as *No Más Bebés*, which chronicles the sterilizations of Mexican and Mexican American women in Los Angeles in the late 1960s. We also really need documentaries like *Baby Bust*, a 2023 Canadian documentary film I participated in that navigates the complexities of climate crisis, eco-anxiety, and the kid question.

Why art? Don't we need political activism to fight climate change? Yes, but the two are not so distinct. Talking to young climate activists has convinced me that a lot of the organizing we see—street demonstrations, strikes, and walkouts—are similar to performance art. These actions are big public dramas; they create a collective experience and evoke shared emotions. Depending on the action, those emotions might be outrage, anger, or inspiration. Any feeling, whether those mentioned or others, like joy and motivation, hope and optimism, can be a powerful spark for political action, whether it's signing a petition, voting, organizing a mass action, or running for office. And it works; the emotional power generated by connecting with others through collective action has real power.

Young climate activists have demonstrated the emotional power in their organizing, in ways both quiet and loud. In 2018, at the age of fifteen, Greta Thunberg started a quiet protest when she went on strike every Friday in front of the Swedish Parliament, refusing to go to school until the government took stronger action to address the climate crisis. That act led to widespread attention and the creation of a global youth movement called Fridays for Future, whereby students do not attend classes on Fridays, opting instead to participate in climate actions. A global student strike for the climate built on Thunberg's school strike in March 2019; there were more than a million participants from 125 countries. Here in the United States, in November 2018, a group of young activists staged a sit-in in Speaker Nancy Pelosi's office, demanding the development of a Green New Deal that would support a just economic transition to 100 percent renewable energy by creating well-paying jobs in the clean energy sector. Brand-new Representative Alexandria Ocasio Cortez, the youngest congresswoman ever elected, joined their protest and helped amplify their strategy, placing it on the national agenda.

Later, Senator Bernie Sanders proposed a Green New Deal policy platform in his run for the presidency in 2020.

There are many impactful campaigns being led by climate justice groups all over the world. The Rise Up Climate Movement, founded by Ugandan climate activist Vanessa Nakate, works to amplify the voices of African climate activists; SustyVibes in Nigeria helps make environmental sustainability relatable and actionable for African youth (the founder, Jennifer Uchendu, is an outspoken advocate for awareness of eco-anxiety in the Global South); meanwhile, the Health and Environment-friendly Youth Campaign, founded by Ashley Lashley, works with young people across the Caribbean to raise awareness of the impacts of climate change on human health and the health of the planet. There are many more grassroots movements arising every day, led by dynamic youth committed to demanding action from governments, corporations, and citizens to address climate change.

Activists like Emma Lim, Blythe Pepino, and Jessica Gaitán Johannesson also strike this chord. Regardless of the critiques of their messaging, there is power in what they have done: they have put their eco-anxieties on public display, for the sake of connecting with others through reproductive resistance. More important, they have generated an important discussion about the relationship between climate crisis and reproductive anxiety—specifically, how the systemic drivers of climate change are causing young people to rethink having children. Even though these discussions attract controversy and inadvertently give space to people who blame population growth for what are actually political and economic problems, they also help reproductively anxious people feel seen and heard. Conceivable Future does this as well, though on a private basis. These discussions resituate private feelings in public context and let people who share these feelings know that they are actually in a bigger, broader community of people who feel similarly. People who would otherwise suffer in silence with their fears and anxieties about climate disruption. People who, through collective conversation with others, might find the confidence to forge ahead and have the families they want.

Earlier in this book, I showed that general attitudes toward having children in the United States have shifted: people are having fewer children, some aren't having any at all, and many people who don't have them are quite happy about it, although they are often judged and stigmatized for it. What's also happening is that the social conditions that generally support parenting—like a decent salary, good health care, child care, paid or unpaid time off work to birth and bond with those kids—are not guaranteed by our social institutions. And climate change impacts, including those in the realm of emotional and mental health challenges, are making it so that bringing children into this mix feels insurmountable.

This is exactly why children are so important. We need to be able to pass on family traditions and cultural knowledge to younger generations. We need to know that our determination to survive and thrive, and leave a meaningful legacy in the future, will be carried out. We need to be able to say, *Some aspect of me, my people, my community will live on.* And we cannot do this in the absence of comprehensive public policies that support families and caregivers—from employment policies that comprehensively cover parent and family leave, to employment and insurance programs that support fertility services, and affordable child care and elder care. We need living wages so that we can work fewer hours and actually spend time with our families without being overwhelmed by stress, and we need government programs that adequately provide these services instead of privatizing the costs and pushing them onto parents', and particularly mothers,' shoulders.

Specifically, we need to recognize that the nuclear family model is failing us all.[1] So is valuing individualism and independence over interdependence. After all, as journalist Anna Louie Sussman reminds us, "Relationships, between us and the natural world, and us and one another, testify to the interdependence that capitalist logic would have us disavow . . . we must find concrete ways to recognize this interdependence and resolve to strengthen it."[2] These failures, of the nuclear family model and the individualist ethic, in the context of a growing climate crisis, are making reproductive anxiety worse. When the state externalizes the costs of creating and sustaining families, the nuclear family buckles under the stress of a weight that it was never intended to support. Seeing that, why would young people want to create nuclear families under the looming, monumental weight of climate change—in addition to the already ongoing crises of racial terror, class warfare, xenophobic hatred, extreme political polarization, the rollback of reproductive autonomy, and the relentless assaults on LGBTQ+ rights?

But there are other models for making family. Marginalized communities have been creating alternative forms of kinship for generations, outside of the recognition of—and often in resistance to—the state that fails them. LGBTQ+ people have long demonstrated that you can make families outside of legal marriage, childbearing, or childrearing, and that challenging the privilege accorded to biological ideas of family is an important step in creating and sustaining chosen kinship.[3] Black families have, since the days of slavery, had no choice but to turn to extended relatives and multiple generations of family members in the presence of social, political, and economic crises and the absence of effective social services. Sociologist Ruha Benjamin notes that Black people have always made these alternative kin relationships, even including ancestors in the mix, as a way of rejecting social abandonment.[4] Indigenous studies scholars note that extended kinship ties and a decentering of marriage and nuclear families have always provided for kin in a variety of ways.[5] Disability justice advocates call for including disabled community members in flexible structures of care and reimagining kinship—again outside of and in response to oppressive nuclear family models.[6] The nuclear family is failing us, but there has always been another way.

Marginalized groups have created these communities of care for generations. Parents and communities of color, as well as queer, trans, and disabled communities, have long collectivized to coordinate meeting their needs in the absence of effective care and support from the government. Mutual aid refers to how groups self-organize to address shared needs through mutual support; it includes things like raising and distributing money, sharing rides, training to deliver medical services, organizing advocacy campaigns, and so on. Marginalized communities provide mutual aid because they recognize that the conditions they are living in are unjust, and that these conditions will never provide the support communities need to survive and thrive.[7]

Mutual aid groups thrived during the early days of the Covid-19 pandemic, garnering mainstream attention as more white and middle-class groups engaged with them, but it is a common long-term practice among vulnerable communities—and will become even more important in the ongoing climate crisis. Supporting individuals and families in being resilient after climate disasters will require these systems of support, outside of the capitalist systems that create and exacerbate climate crisis. This is how marginalized communities have sustained themselves through ongoing existential crises, from land theft and the legacies of racial terror including genocide and slavery, to the everyday social, political, and medical violence enacted against queer and trans people. While mutual aid may seem like a collection of instrumental resources, it offers a boon to mental and emotional health and well-being.[8]

I have two goals for this book. The first is that it will help those who are struggling with climate anxiety and the kid question to feel seen. I want to remove the shroud of silence from what is actually a common experience for a lot of climate-aware people in their reproductive years. Related to that, this is a complicated issue that people experience differently based on race and class, and we really need to center the voices of marginalized people of color because those voices are all too often silenced, although they have much to say about being resilient in the face of threats to survival. In addition, culturally competent emotional and mental health resources are desperately needed in communities of color, where they are all too often lacking. This is as much a climate justice issue as polluted air or toxic soil.

That brings me to my second goal for the book. I hope it will help serve as the impetus for more research on environmental and climate emotions among marginalized communities, including communities or color as well as queer and trans communities. We need more of the quantitative studies that can tell us what large numbers of people are thinking, feeling, and doing. We need the qualitative research that asks the why question and goes in-depth to find out the answers. More important, we need research that does not assume that climate emotions are experienced the same way everywhere, by everyone—or, worse, that marginalized people don't experience these emotional impacts at all. The exclusion of people of color from research on climate emotions and reproductive plans prevents us from knowing more about how interpersonal racism and systematic racial inequality permeate how climate impacts land on the body and the emotions, as well as how they pervade the most intimate questions about how we plan our lives and families. We also need climate emotions research that prioritizes LGBTQ+ communities, who are systematically left out of reporting on climate disasters and experience discrimination in disaster recovery and relief efforts.[9] Excluding marginalized communities from research studies constitutes an injustice, because the lack of data makes it difficult to advocate for policies and resources needed to address the differential burdens we experience. The research is necessary to establish an evidence base. From there, we need a more comprehensive climate justice approach that includes addressing environmental and climate emotions, including accessible and culturally competent mental health services.[10]

With that said, I want to resist the impulse to reify social problems as individual mental disorders. To paraphrase scholar Danielle Carr, climate change is a crisis that *affects* mental health, rather than a crisis *of* mental health.[11] As such, we need to treat mental and emotional suffering, and—most important—we need to address the roots of that suffering. We need mental and emotional health interventions that prioritize marginalized communities hit hardest by climate impacts, and we need these services integrated into broader efforts to achieve climate justice moving forward. In this way, and with strengthened, resilient families and communities, the future can actually be one that more of us can look forward to.

Glossary

antinatalism	Philosophical and social stance against procreation
(Baby) Boomer	Those born between 1946 and 1964
childless	A person who does not have a child, whether by choice or circumstance
childfree	A lifestyle and/or movement celebrating the intentional decision not to birth or raise a child
climate anxiety	Emotional and mental responses to anthropogenic climate change
climate justice	Social movement seeking to redress the differential and unequal burdens of climate impacts on vulnerable communities
eco-anxiety	Multifaceted emotional and mental response to ecological and environmental changes
ecofascism	White supremacist ideology that blames immigration and "overpopulation" for environmental problems
Generation X	Those born between 1965 and 1980
Generation Z	Those born between 1997 and 2012

kid question	The issue of whether, when, and how to have a child while grappling with climate crisis
Millennial	Those born between 1981 and 1996
neo-Malthusian	Perspective that identifies population growth as the primary cause of environmental problems
reproductive anxiety	Mental and emotional reaction to the moral, ethical, and social concerns one has about birthing and raising children during the climate crisis
Reproductive Justice	Social movement advocating the right to not have unwanted children, have wanted children, and raise children in safe and sustainable environments
reproductive refusal	An active and ongoing set of decisions to not birth or raise children
reproductive resilience	A commitment to birthing and/or raising children while navigating large-scale structural challenges
reproductive resistance	An active way of fighting back against the undesirable conditions shaping pregnancy, birth, and parenting

Notes

1. I use the term "women" throughout the book for two reasons: first, because of the specific pressures and expectations around pregnancy and motherhood that women and girls are socialized into from birth; and second, because that is how my interviewees and survey participants identify themselves. With that said, I acknowledge that reproductive autonomy is under threat for all people who have the capacity to become pregnant and that societal and family pressure to have children is often applied to anyone with a uterus regardless of gender identity.

INTRODUCTION

1. Interviewee names are pseudonyms.
2. American Psychiatric Association, "APA Public Opinion Poll-Annual Meeting 2020," www.psychiatry.org/newsroom/apa-public-opinion-poll-2020 (accessed July 2022).
3. Marie Haaland, "Majority of Young American Adults Say Climate Change Influences Their Decision to Have Children," *SWNS Digital*,

September 6, 2021, https://swnsdigital.com/us/2020/04/majority-of-young-american-adults-say-climate-change-influences-their-decision-to-have-children/.

4. Centers for Disease Control and Prevention, "Climate Effects on Health," www.cdc.gov/climateandhealth/effects/default.htm (accessed April 2023).

5. Intergovernmental Panel on Climate Change (IPCC), "Summary for Policymakers," in *Global Warming of 1.5°C. An IPCC Special Report on the Impacts of Global Warming of 1.5°C above Pre-Industrial Levels and Related Global Greenhouse Gas Emission Pathways, in the Context of Strengthening the Global Response to the Threat of Climate Change, Sustainable Development, and Efforts to Eradicate Poverty,* ed. V. P. Masson-Delmotte, P. Zhai, H.-O. Pörtner, et al. (IPCC, 2018), www.ipcc.ch/sr15/chapter/spm/ (accessed August 2022).

6. Ella Nilsen, "Biden Administration Approves Controversial Willow Oil Project in Alaska, Which Has Galvanized Online Activism," *CNN*, March 13, 2023, www.cnn.com/2023/03/13/politics/willow-project-alaska-oil-biden-approval-climate/index.html.

7. Environmental Protection Agency (EPA), *Climate Change and Social Vulnerability in the United States: A Focus on Six Impacts*, EPA 430-R-21-003, 2021, www.epa.gov/cira/social-vulnerability-report (accessed May 2022). Also see Rachel Morello-Frosch, Manuel Pastor, James Sadd, and Seth B. Shonkoff, "The Climate Gap: Inequalities in How Climate Change Hurts Americans and How to Close the Gap," University of Southern California, 2009, https://dornsife.usc.edu/assets/sites/242/docs/ClimateGapReport_full_report_web.pdf (accessed August 2022).

8. Sarah Jaquette Ray, "Climate Anxiety Is an Overwhelmingly White Phenomenon," *Scientific American*, March 21, 2021, www.scientificamerican.com/article/the-unbearable-whiteness-of-climate-anxiety/#.

9. Mary Annaïse Heglar, "Climate Change Isn't the First Existential Threat," *Medium*, February 18, 2019, https://zora.medium.com/sorry-yall-but-climate-change-ain-t-the-first-existential-threat-b3c999267aa0.

10. Britt Wray, *Generation Dread: Finding Purpose in an Age of Climate Crisis* (New York: Alfred A. Knopf, 2022), 4.

11. Morning Consult, "National Tracking Poll #200926: Crosstabulation Results," September 8–10, 2020.

12. Matthew Ballew, Edward Maibach, John Kotcher, et al., "Which Racial/Ethnic Groups Care Most about Climate Change?" Yale Program on

Climate Change Communication, 2020, https://climatecommunication .yale.edu/publications/race-and-climate-change/ (accessed June 2022).

13. Sri S. Uppalapati, Matthew Ballew, Eryn Campbell, John Kotcher, Seth Rosenthal, Anthony Leiserowitz, and Edward Maibach, "The Prevalence of Climate Change Psychological Distress among American adults," Yale Program on Climate Change Communication, 2023, https://climate communication.yale.edu/publications/climate-change-psychological-distress-prevalence/ (accessed September 23, 2023).

14. Jade Sasser, *On Infertile Ground: Population Control and Women's Rights in the Era of Climate Change* (New York: NYU Press, 2018).

15. Kristy Campion, "Defining Ecofascism: Historical Foundations and Contemporary Interpretations in the Extreme Right," *Terrorism and Political Violence* 35, no. 4 (2023): 926–944. DOI: 10.1080/09546553.2021.1987895.

16. Daniel Sherrell, *Warmth: Coming of Age at the End of Our World* (New York: Penguin Books, 2021), 18.

17. Glenn Albrecht, "'Solastalgia' a New Concept in Health and Identity," *Philosophy Activism Nature* 3 (2005): 41–55.

18. Glenn Albrecht, "Chronic Environmental Change: Emerging 'Psychoterratic' Syndromes," chapter 3 in *Climate Change and Human Well-Being. International and Cultural Psychology*, ed. Inka Weissbecker (New York: Springer, 2011), https://doi.org/10.1007/978-1-4419-9742-5_3.

19. Renee Lertzman, *Environmental Melancholia: Psychoanalytic Dimensions of Engagement* (London: Routledge, 2015).

20. Ashlee Cunsolo and Neville R. Ellis, "Ecological Grief as a Mental Health Response to Climate Change-Related Loss," *Nature Climate Change* 8 (2018): 275–281.

21. Naomi Klein, *This Changes Everything: Capitalism vs. the Climate* (New York: Simon & Schuster, 2014), 420.

22. Thomas Doherty and Susan Clayton, "The Psychological Impacts of Global Climate Change," *American Psychologist* 66, no. 4 (2011): 265–276.

23. David Theo Goldberg, *Dread: Facing Futureless Futures* (Cambridge, UK: Polity Press, 2021), 15.

24. Leo Goldsmith, Vanessa Raditz, and Michael Méndez, "Queer and Present Danger: Understanding the Disparate Impacts of Disasters on LGBTQ+ Communities," *Disasters* 46, no. 4 (2022): 946–973; Michael Méndez, Genevieve Flores-Haro, and Lucas Zucker, "The (in)visible Victims of Disaster: Understanding the Vulnerability of Undocumented Latino/a and Indigenous Immigrants," *Geoforum* 116 (2020): 50–62.

25. Farhana Sultana, "The Unbearable Heaviness of Climate Coloniality," *Political Geography*, 99 (2022): 102638, https://doi.org/10.1016/j.polgeo.2022.102638.

26. Panu Pihkala, "Toward a Taxonomy of Climate Emotions," *Frontiers in Climate* 3 (2022): 738154, DOI: 10.3389/fclim.2021.738154.

27. Panu Pihkala, "Anxiety and the Ecological Crisis: An Analysis of Eco-Anxiety and Climate Anxiety," *Sustainability* 12, no. 19 (2020): 7836, https://doi.org/10.3390/su12197836.

28. Panu Pihkala, *Climate Anxiety* (Helsinki: Mieli Mental Health Finland, 2019).

29. Pihkala, "Anxiety and the Ecological Crisis," 14.

30. Caroline Hickman, "We Need to (Find a Way to) Talk about . . . Eco-Anxiety.," *Journal of Social Work Practice* 34, no. 4 (2020): 411–424, 414–415.

31. For example, see David M. Abramson, Yoon Soo Park, Tasha Stehling-Ariza, and Irwin Redlener, "Children as Bellwethers of Recovery: Dysfunctional Systems and the Effects of Parents, Households, and Neighborhoods on Serious Emotional Disturbance in Children after Hurricane Katrina," *Journal of Disaster Medicine and Public Health Preparedness*, 4 Suppl. 1: S17-S27 (2010). Elizabeth Fussell and Sarah R. Lowe, "The Impact of Housing Displacement on the Mental Health of Low-Income Parents after Hurricane Katrina," *Social Science and Medicine* 113 (2014): 137–144.

32. Adam C. Alexander, Jeanelle Ali, Meghan E. McDevitt-Murphy, David R. Forde, Michelle Stockton, Mary Read, and Kenneth D. Ward, "Racial Differences in Posttraumatic Stress Disorder Vulnerability Following Hurricane Katrina among a Sample of Adult Cigarette Smokers from New Orleans," *Journal of Racial and Ethnic Health Disparities* 4, no. (2017): 94–103, DOI:10.1007/s40615-015-0206-8.

33. Charles Ogunbode et al., "Negative Emotions about Climate Change Are Related to Insomnia Symptoms and Mental Health: Cross Sectional Evidence from 25 Countries," *Current Psychology* (2021), https://doi.org/10.1007/s12144-021-01385-4.

34. Charles Ogunbode et al., "Climate Anxiety, Pro-Environmental Action and Wellbeing: Antecedents and Outcomes of Negative Emotional Responses to Climate Change in 28 Countries," *Journal of Environmental Psychology* 84 (2022): 101887.

35. Caroline Hickman, Elizabeth Marks, Panu Pihkala, Susan Clayton, Eric R. Lewandowski, Elouise E. Mayall, Britt Wray, Catriona Mellor, and

Lise van Susteren, "Climate Anxiety in Children and Young People and Their Beliefs about Government Responses to Climate Change: A Global Survey," *The Lancet Planetary Health* 5, no. 12 (2021): e863-e873.

36. Climate Critical Earth, "Climate Burnout Report: An Exploratory Study of Burnout in the Climate and Environmental Workforce," www .climatecritical.earth/report (accessed May 2023).

37. Hickman, "We Need to (Find a Way to) Talk About."

38. Rebecca Solnit, "Difficult Is Not the Same as Impossible," in *Not Too Late: Changing the Climate Story from Despair to Possibility*, ed. Rebecca Solnit and Thelma Young Lutunatabua (Chicago: Haymarket Books, 2023), 6.

39. Mary Annaïse Heglar, "Here's Where You Come In," in *Not Too Late*, ed., Solnit and Lutunatabua, 25.

40. Meehan Crist, "Is It OK to Have a Child?" *London Review of Books* 42, no. 5 (March 5, 2020), www.lrb.co.uk/the-paper/v42/n05/meehan-crist /is-it-ok-to-have-a-child.

41. Thelma Young Lutunatabua, "Not Only a Danger but a Promise," in *Not Too Late*, ed. Solnit and Lutunatabua, 197.

CHAPTER 1

1. Molly Lambert, "Miley Cyrus Has Finally Found Herself," *Elle*, July 11, 2019, www.elle.com/culture/music/a28280119/miley-cyrus-elle-interview/.

2. Elle Hunt, "Birthstrikers: Meet the Women Who Refuse to Have Children until Climate Change Ends," *The Guardian*, March 12, 2019, www .theguardian.com/lifeandstyle/2019/mar/12/birthstrikers-meet-the-women-who-refuse-to-have-children-until-climate-change-ends?fbclid=IwAR31PA5 Q1ixnbnB9cj3JAENb_XlmnT-vhS-VDQUbYUg-THg7AdIO7TD88zE.

3. Sam Shead, "Climate Change Is Making People Think Twice about Having Children," CNBC, August 13, 2021, www.cnbc.com/2021/08 /12/climate-change-is-making-people-think-twice-about-having-children .html.

4. Amrita Chakrabarti Myers, "'Sisters in Arms': Slave Women's Resistance to Slavery in the United States," *Past Imperfect* 5 (1966): 141–174.

5. Jane Recker, "When Abortion Was Illegal, Chicago Women Turned to the Jane Collective," *Smithsonian Magazine*, June 14, 2022, www.smith sonianmag.com/smart-news/abortion-jane-collective-chicago-180980244/.

6. Joyce C. Follett, *Making Democracy Real: African American Women, Birth Control, and Social Justice, 1910–1960: The Negro Project*, 2019, Sofia Smith Collection, Smith College, https://sophia.smith.edu/making-democracy-real/the-negro-project/ (accessed August 1, 2022).

7. Gretchen Livingston and D'vera Cohn, "Childlessness Up among All Women; Down among Women with Advanced Degrees," Pew Research Center, 2010, www.pewresearch.org/social-trends/2010/06/25/childlessness-up-among-all-women-down-among-women-with-advanced-degrees/ (accessed July 2022).

8. Anna Brown, "Growing Share of Childless Adults in U.S. Don't Expect to Ever Have Children," Pew Research Center, 2021, www.pewresearch.org/fact-tank/2021/11/19/growing-share-of-childless-adults-in-u-s-dont-expect-to-ever-have-children/ (accessed July 2022).

9. Department of Health and Human Services, Administration for Children and Families, *Trends in Foster Care and Adoption: FY 2012–21*, www.acf.hhs.gov/cb/report/trends-foster-care-adoption (accessed May 2023); also see E. Koh, R. Hanlon, L. Daughtery, et al., *Adoption by the Numbers* (Alexandria, VA: National Council for Adoption, 2022).

10. Amy Blackstone, *Childfree by Choice: The Movement Redefining Family and Creating a New Age of Independence* (New York: Penguin Random House, 2019).

11. Carolyn Morell, "Saying No: Women's Experiences with Reproductive Refusal," *Feminism & Psychology* 10, no. 3 (2000): 313–322.

12. Leslie Ashburn-Nardo, "Parenthood as a Moral Imperative? Moral Outrage and the Stigmatization of Voluntarily Childfree Women and Men," *Sex Roles* 76 (2016): 393–401.

13. See a list of OECD countries at www.oecd.org/about/document/ratification-oecd-convention.htm.

14. Jennifer Glass, Robin Simon, and Matthew Anderson, "Parenthood and Happiness: Effects of Work-Family Reconciliation Policies in 22 OECD Countries," *American Journal of Sociology* 122, no. 3 (2016): 886–929.

15. Daniel Kahneman, Alan B. Krueger, David Schkade, Norbert Schwarz, and Arthur Stone, "Toward National Well-being Accounts," *American Economic Review* 94, no. 2 (2004): 429–434.

16. Angus Deaton and Arthur A. Stone, "Evaluative and Hedonic Well-being among Those with and without Children at Home," *PNAS* 111, no. 4 (2014): 1328–1333.

17. The NotMom Summit, www.thenotmom.com/notmom-summit (accessed August 1, 2022).

18. Carolyn Morell, "Saying No: Women's Experiences with Reproductive Refusal," *Feminism & Psychology* 10, no. 3 (2000): 313–322.

19. Amy Blackstone, *Childfree by Choice: The Movement Redefining Family and Creating a New Age of Independence* (New York: Penguin Random House, 2019), 3–4.

20. Blackstone, *Childfree by Choice*, 18.

21. Rachel G. Riskind and Charlotte J. Patterson, "Parenting Intentions and Desires among Childless Lesbian, Gay, and Heterosexual Individuals," *Journal of Family Psychology* 24 (2010): 78–81; and Anthony R. D'Augelli, H. Jonathon Rendina, Katerina O. Sinclair, and Arnold H. Grossman, "Lesbian and Gay Youth's Aspirations for Marriage and Raising Children," *Journal of LGBT Issues in Counseling* 1 (2007): 77–98.

22. Mignon Moore, *Invisible Families: Gay Identities, Relationships, and Motherhood among Black Women* (Berkeley: University of California Press, 2011).

23. Nikki Hayfield, Gareth Terry, Victoria Clarke, and Sonja Ellis, "'Never Say Never?' Heterosexual, Bisexual, and Lesbian Women's Accounts of Being Childfree," *Psychology of Women Quarterly* 43, no. 4 (2019): 526–538.

24. Britt Wray, *Generation Dread: Finding Purpose in an Age of Climate Crisis* (Toronto: Alfred A. Knopf, 2022).

25. Sacoby M. Wilson, "Roundtable on the Pandemics of Racism, Environmental Injustice, and Covid-19 in America," *Environmental Justice* 13, no. 3 (2020): 56–64.

26. Caroline Hickman, Elizabeth Marks, Panu Pihkala, Susan Clayton, Eric R. Lewandowski, Elouise E. Mayall, Britt Wray, Catriona Mellor, and Lise van Susteren, "Climate Anxiety in Children and Young People and Their Beliefs about Government Responses to Climate Change: A Global Survey," *Lancet Planetary Health* 5, no. 12 (2021): e863–e873.

27. Blue Shield of California, *Survey Report: Nextgen Climate Survey*, 2021 (accessed July 15, 2022).

28. Climate Mental Health Network, "How to Talk to Young People about Climate Emotions," https://climatementalhealth.net/parents (accessed May 1, 2023).

29. Larissa Dooley, Jylana Sheats, Olivia Hamilton, Dan Chapman, and Beth Karlin, *Climate Change and Youth Mental Health: Psychological Impacts, Resilience Resources, and Future Directions* (Los Angeles: See Change Institute, 2021).

CHAPTER 2

1. Rickie Solinger, *Pregnancy and Power: A Short History of Reproductive Politics in America* (New York: NYU Press, 2005).

2. Loretta Ross and Rickie Solinger, *Reproductive Justice: An Introduction* (Oakland: University of California Press, 2017).

3. Sabrina Tavernise, "The U.S. Birthrate Has Dropped Again. The Pandemic May Be Accelerating the Decline," *New York Times*, May 5, 2021, www.nytimes.com/2021/05/05/us/us-birthrate-falls-covid.html.

4. Luke Rogers, "Covid-19, Declining Birth Rates, and International Migration Resulted in Historically Small Population Gains," US Census Bureau, 2021, www.census.gov/library/stories/2021/12/us-population-grew-in-2021-slowest-rate-since-founding-of-the-nation.html (accessed July 29, 2022).

5. Derek Thompson, "Why U.S. Population Growth Is Collapsing," *The Atlantic*, March 28, 2022, www.theatlantic.com/newsletters/archive/2022/03/american-population-growth-rate-slow/629392/.

6. Thompson, "Why U.S. Population Growth Is Collapsing."

7. Shawn Hubler, "Why California's Growth Has Slowed (and Why Demographers Aren't Surprised)," *New York Times*, May 10, 2021, www.nytimes.com/2021/04/26/us/us-census-california.html.

8. S. E. Vollset, E. Goren, C-W Yuan, J. Cao, et al., "Fertility, Mortality, Migration, and Population Scenarios for 195 Countries and Territories from 2017 to 2100: A Forecasting Analysis for the Global Burden of Disease Study," *Global Health Metrics* 396 (2020): 1285–1306.

9. P. Friedlingstein, M. O'Sullivan, M. W. Jones, R. Andrew, et al., "Global Carbon Budget 2020," *Earth System Science Data* 12, no. 4 (2020): 3269–3340, https://essd.copernicus.org/articles/12/3269/2020/.

10. C. Le Quere, R. B. Jackson, M. W. Jones, A. Smith, et al., "Temporary Reduction in Daily Global CO_2 Emissions during the Covid-19 Forced Confinement," *Nature Climate Change* 10 (2020): 647–653, www.nature.com/articles/s41558-020-0797-x.

11. Cassidy Thomas and Elhom Gosink, "At the Intersection of Eco-Crises, Eco-Anxiety, and Political Turbulence: A Primer on Twenty-First Century Ecofascism," *Perspectives on Global Development and Technology* 20 (2021): 30–54; and Jason Wilson and Aaron Flanagan, "The Racist 'Great Replacement' Conspiracy Theory Explained," *Southern Poverty Law Center*, May 17, 2022, www.splcenter.org/hatewatch/2022/05/17/racist-great-replacement-conspiracy-theory-explained.

12. Steven L. Myers, Jin Wu, and Clare Fu, "China's Looming Crisis: A Shrinking Population," *New York Times*, January 17, 2020, www.nytimes .com/interactive/2019/01/17/world/asia/china-population-crisis.html.

13. Heather Chen, "Forget Tinder. This Chinese City Is Building a Dating Database for Its Singles," *Vice*, December 28, 2021, www.vice.com/en /article/akvkwe/chinese-city-building-dating-database-for-singles.

14. Jenny Brown, *Birth Strike: The Hidden Fight over Women's Work* (Oakland, CA: PM Press, 2019).

15. Damien Cave, Emma Bubola, and Choe Sang-Hun, "Long Slide Looms for World Population, with Sweeping Ramifications," *New York Times*, May 22, 2021, www.nytimes.com/2021/05/22/world/global-population-shrinking.html.

16. Spies Rejser, "Do It for Denmark!" YouTube, 2014, www.youtube .com/watch?v = vrO3TfJc9Qw (accessed June 2022).

17. Danica Jefferies, JoElla Carman, and Nigel Chiwaya, NBC News, "Abortion Law Tracker: See Where the Procedure Is Currently Legal, Banned or Restricted in the US," June 2022, updated August 23, 2023, www.nbcnews.com/data-graphics/abortion-state-tracking-trigger-laws-bans-restrictions-rcna36199; and "Tracking the States Where Abortion Is Now Banned," *New York Times*, May 5, 2023, updated September 19, 2023 www.nytimes.com/interactive/2022/us/abortion-laws-roe-v-wade.html.

18. Brown, *Birth Strike*.

19. Brown, *Birth Strike*, 11.

20. Laura Briggs, *How All Politics Became Reproductive Politics: From Welfare Reform to Foreclosure to Trump* (Oakland: University of California Press, 2018).

21. Caitlin Collins, *Making Motherhood Work: How Women Manage Careers and Caregiving* (Princeton, NJ: Princeton University Press, 2019), 2.

22. Collins, *Making Motherhood Work*.

23. Rochelle Green, Varada Sarovar, Brian Malig, and Rupa Basu, "Association of Stillbirth with Ambient Air Pollution in a California Cohort Study," *American Journal of Epidemiology* 181, no. 11 (2015): 874–882; and Rupa Basu, Darshani Pearson, Keita Ebisu, and Brian Malig, "Association between PM2.5 and PM2.5 Constituents and Preterm Delivery in California, 2000– 2006," *Pediatric Perinatal Epidemiology* 31, no. 5 (2017): 424–434

24. CDC, *Reproductive Health: Preterm Birth*, 2021, www.cdc.gov /reproductivehealth/maternalinfanthealth/pretermbirth.htm (accessed May 2022).

25. Danielle Renwick, "The U.S. Is Facing a Maternal Health Crisis. Climate Change Is Making It Worse," *The Nation*, March 17, 2022, www.thenation.com/article/environment/climate-change-maternal-health/.

26. Alan Barreca and Jessamyn Schaller, "The Impact of High Ambient Temperatures on Delivery Timing and Gestational Lengths," *Nature Climate Change* 10 (2019): 77–82.

27. Rupa Basu, Reina Rau, Darshani Pearson, and Brian Malig, "Temperature and Term Low Birth Weight in California," *American Journal of Epidemiology* 187, no. 11 (2018): 2306–2314.

28. Zahra Hirji, "The Coronavirus Pandemic Is Making the Threat of Summer Heat Worse for Pregnant Women," *Buzzfeed News*, August 1, 2020, www.buzzfeednews.com/article/zahrahirji/coronavirus-pregnant-health-heat.

29. Anna M. Phillips, Tony Barboza, Ruben Vives, and Sean Greene, "Heat Waves Are Far Deadlier Than We Think. How California Neglects this Climate Threat," *LA Times*, October 7, 2021, www.latimes.com/projects/california-extreme-heat-deaths-show-climate-change-risks/#nt = 0000017c-3247-d42d-adfd-32ef494d0000-showMedia-title-promo SuperLeadSmall-1col-enhancement.

30. "Booker, Underwood, Adams, Senators Unveil Black Maternal Health Momnibus Act to Address America's Maternal Health Crisis," Office of Senator Cory Booker, February 8, 2021, www.booker.senate.gov/news/press/booker-underwood-adams-senators-unveil-black-maternal-health-momnibus-act-to-address-americas-maternal-health-crisis.

31. US Congress, "S.423-Protecting Moms and Babies Against Climate Change Act," www.congress.gov/bill/117th-congress/senate-bill/423 (accessed May 2022).

32. Laura D. Lindberg, Jennifer Mueller, Marielle Kirstein, and Alicia VandeVusse, *The Continuing Impacts of the COVID-19 Pandemic in the United States: Findings from the 2021 Guttmacher Survey of Reproductive Health Experiences* (New York: Guttmacher Institute, 2021), www.guttmacher.org/report/continuing-impacts-covid-19-pandemic-findings-2021-guttmacher-survey-reproductive-health. https://doi.org/10.1363/2021.33301.

33. Christina Rexrode, "They Were Planning to Get Pregnant. The Coronavirus Complicates Things," *Wall Street Journal*, May 5, 2020, www.wsj.com/articles/they-were-planning-to-get-pregnant-the-coronavirus-makes-that-complicated-11588680007; Sacha Pfeiffer, "Women 'Falling off the Cliff of Fertility' As Pandemic Puts Treatments on Hold," *NPR*, April 12,

2020, www.npr.org/2020/04/12/831706118/women-say-they-are-falling-off-the-cliff-of-fertility-as-pandemic-puts-treatment; and Zeng Hui, Xu Chen, Fan Junli, et al., "Antibodies in Infants Born to Mothers with COVID-19 Pneumonia," *JAMA* 323, no. 18 (2020), DOI:10.1001/jama.2020.4861.

34. Renjen Punit, "Women Are Vanishing from the Workplace—Here's How to Bring Them Back," *The Hill*, June 13, 2021, https://thehill.com/opinion/campaign/558180-women-are-vanishing-from-the-workplace-heres-how-to-bring-them-back/.

35. Matt Krentz, Emily Kos, Anna Green, and Jennifer Garcia-Alonso, "Easing the Covid-19 Burden on Working Parents," *Boston Consulting Group*, May 21, 2020, www.bcg.com/publications/2020/helping-working-parents-ease-the-burden-of-covid-19; and A. Adams-Prassl, T. Boneva, M. Golin, and C. Rauh, *Inequality in the Impact of the Coronavirus Shock: Evidence from Real Time Surveys*, 2020, Cambridge-INET Working Paper WP2018, DOI:10.17863/CAM.57992.

36. Victoria Reyes, *Academic Outsider: Stories of Exclusion and Hope* (Stanford, CA: Stanford University Press, 2022), 87.

CHAPTER 3

1. Claire Cain Miller, "Americans Are Having Fewer Babies. They Told Us Why," *New York Times*, July 5, 2018, www.nytimes.com/2018/07/05/upshot/americans-are-having-fewer-babies-they-told-us-why.html.

2. Reina Pomeroy, "Is Climate Change a Factor in Fertility Decisions? TL; DR: For Many, Yes," ModernFertility.com, December 1, 2021, https://modernfertility.com/blog/climate-change-and-fertility-survey/?clickid = S4CUePzk%3AzZDWofyqbyQSXKPUkDz6OX1p1O4xI0&utm_source = Impact&utm_medium = 10079&utm_campaign = Online%20Tracking%20Link&utm_content = Skimbit%2C%20Ltd.&irgwc = 1.

3. Marie Haaland, "Majority of Young American Adults Say Climate Change Influences Their Decision to Have Children," SWNS Digital, September 6, 2021, https://swnsdigital.com/us/2020/04/majority-of-young-american-adults-say-climate-change-influences-their-decision-to-have-children/.

4. Morning Consult, "National Tracking Poll #200926: Crosstabulation Results," September 8–10, 2020.

5. Statista Research Department, "Age of Mothers at First Birth in the U.S. by Hispanic Origin, 2020," Statista, 2022, www.statista.com

/statistics/260386/mean-age-of-mothers-at-first-birth-in-the-united-states-in-by-hispanic-origin/ (accessed April 1, 2023).

6. In terms of race, just under half were non-Hispanic whites; 20 percent were Black or African American, 19 percent were Hispanic or Latino; just under 10 percent were Asian; and the remainder identified as mixed race or other. Half identified as men, half as women (fewer than 1 percent identified as nonbinary); 20 percent were between the ages of 22 and 25, 40 percent were 26 to 30, and the remaining 40 percent were 31 to 35; 44 percent had high school degrees only; 17 percent had associates degrees, 29 percent had bachelor's, and the remaining 10 percent had graduate degrees. Half were Christian. One-third had no religion at all. Politically, most responses came from people right in the middle: moderate centrists. While 21 percent identified as liberal and 13 percent were very liberal, 15 percent were apolitical, 12 percent identified as conservative, and 5 percent were very conservative. A full third, 33 percent, live in the South or Southeast; 24 percent in the Northeast; 20 percent in the Midwest. My own region, the West/Southwest, had the smallest proportion at 15 percent. And finally, in terms of class, most people identified as either middle income (their households earn between $40,000 and $120,000 per year) or working class or lower income (less than $40,000 per year). Just 8 percent live in households that earn more than $10,000 per year.

7. Kaiser Family Foundation, "Deaths Due to Firearms per 100,000 Population by Race/Ethnicity," 2020, www.kff.org/other/state-indicator/firearms-death-rate-by-raceethnicity/?currentTimeframe = 0&print = true&selectedRows = %7B%22wrapups%22:%7B%22united-states%22:%7B%7D%7D%7D&sortModel = %7B%22colId%22:%22Location%22,%22sort%22:%22asc%22%7D (accessed May 1, 2023); and Gabriel L. Schwartz and Jaquelyn L. Jahn, "Mapping Fatal Police Violence across U.S. Metropolitan Areas: Overall Rates and Racial/Ethnic Inequities, 2013–2017," *PLOS One* (2020): https://doi.org/10.1371/journal.pone.0229686.

8. Native People's Law Project, *Native Lives Matter*, 2015, https://lakota-prod.s3-us-west-2.amazonaws.com/uploads/Native-Lives-Matter-PDF.pdf (accessed May 1, 2023).

9. Annita Lucchesi and Abigail Echo-Hawk, *Missing and Murdered Indigenous Women & Girls: A Snapshot of Data from 71 Urban Cities in the United States,* Urban Indian Health Institute, 2016, www.uihi.org/wp-content/uploads/2018/11/Missing-and-Murdered-Indigenous-Women-and-Girls-Report.pdf (accessed August 1, 2022).

10. Seattle Indian Health Board, *Community Health Profile: National Aggregate of Urban Indian Health Program Service Areas*, Urban Indian Health Institute, 2016 (accessed May 1, 2023).

11. Rachel Morello-Frosch, Manuel Pastor, James Sadd, and Seth Shinkoff, "The Climate Gap: Inequalities in How Climate Change Hurts Americans & How to Close the Gap," USC Dornsife, 2009, https://dornsife.usc.edu/eri/the-climate-gap-inequalities-how-climate-change-hurts-america/ (accessed June 1, 2022).

12. US Environmental Protection Agency, *Climate Change and Social Vulnerability in the United States: A Focus on Six Impacts*, EPA 430-R-21-003, www.epa.gov/cira/social-vulnerability-report (accessed June 2022).

13. Morello-Frosch et al., "Climate Gap."

14. US EPA, *Climate Change and Social Vulnerability in the United States*.

15. Susan Clayton, Christie Manning, Meighen Speiser, and Alison Nicole Hill, *Mental Health and Our Changing Climate: Impacts, Inequities, Responses* (Washington, DC: American Psychological Association, and ecoAmerica, 2021).

16. Matthew Ballew, Edward Maibach, John Kotcher, Parrish Bergquist, Seth Rosenthal, Jennifer Marlon, and Anthony Leiserowitz, "Which Racial/Ethnic Groups Care Most about Climate Change?," Yale Program on Climate Change Communication, 2020, https://climatecommunication.yale.edu/publications/race-and-climate-change/ (accessed June 2022).

17. Sri S. Uppalapati, Matthew Ballew, Eryn Campbell, John Kotcher, Seth Rosenthal, Anthony Leiserowitz, and Edward Maibach, "The Prevalence of Climate Change Psychological Distress among American Adults," Yale Program on Climate Change Communication, 2023, https://climatecommunication.yale.edu/publications/climate-change-psychological-distress-prevalence/ (accessed September 23, 2023).

18. Morello-Frosch et al., "Climate Gap."

19. Jade Sasser, *On Infertile Ground: Population Control and Women's Rights in the Era of Climate Chang* (New York: NYU Press, 2018).

20. Steve Arnocky, Darcy Dupuis, and Mirella L. Stroink, "Environmental Concern and Fertility Intentions among Canadian University Students," *Population & Environment* 34 (2012): 279–292.

21. Mark W. Anderson, "The New Ecological Paradigm (NEP) Scale," Berkshire Publishing Group, 2012, https://umaine.edu/soe/wp-content/uploads/sites/199/2013/01/NewEcologicalParadigmNEPScale1.pdf (accessed June 2022).

22. Alessandra De Rose and Maria Rita Testa, "Climate Change and Reproductive Intentions in Europe," Vienna Institute of Demography Working Papers No. 9, 2013, Austrian Academy of Sciences (ÖAW), Vienna Institute of Demography (VID), Vienna.

23. Matthew Schneider-Mayerson and Kit Ling Leong, "Eco-repro Concerns in the Age of Climate Change," *Climatic Change* 163 (2020): 1007–1023.

24. Sabrina Helm, Joya Kemper, and Samantha White, "No Future, No Kids-No Kids, No Future?: An Exploration of Motivations to Remain Childfree in Times of Climate Change," *Population and Environment* 43 (2021): 108–129.

25. Helm, Kemper, and White, "No Future, No Kids-No Kids, No Future?," 118.

26. Leo Goldsmith, Vanessa Raditz, and Michael Méndez, "Queer and Present Danger: Understanding the Disparate Impacts of Disasters on LGBTQ+ Communities," *Disasters* 46, no. 4 (2022): 946–973.

CHAPTER 4

1. Interviewees were between ages twenty-two and thirty-eight; most are from California, although others were from or live in Georgia, Louisiana, and Maryland.

2. Rebecca Hersher and Robert Benincasa, "How Federal Disaster Money Favors the Rich," NPR, March 15, 2019, www.npr.org/2019/03/05/688786177/how-federal-disaster-money-favors-the-rich.

3. Junia Howell and James R. Elliott, "As Disaster Costs Rise, So Does Inequality," *Socius* 4 (2018): 1–3, DOI:10.1177/2378023118816795.

4. Harriet E. Ingle and Michael Mikulewicz, "Mental Health and Climate Change: Tackling Invisible Injustice," *Lancet Planetary Health* 4 (2020): E128–E130.

5. Sarah Jaquette Ray, "Climate Anxiety Is an Overwhelmingly White Phenomenon," *Scientific American*, March 21, 2021, www.scientificamerican.com/article/the-unbearable-whiteness-of-climate-anxiety/.

6. For example, see Todd McCardle, "A Critical Historical Examination of Tracking As a Method for Maintaining Racial Segregation," *Intersectionality and the History of Education* 45 (2020): 2, https://files.eric.ed.gov/fulltext/EJ1250375.pdf.

7. See American Society for Reproductive Medicine (ASRM), "Access to Fertility Services by Transgender and Nonbinary Persons: An Ethics Com-

mittee Opinion," *Fertility & Sterility* 115 (2021): 874–878, www.asrm.org /globalassets/asrm/asrm-content/news-and-publications/ethics-committee-opinions/access_to_care_for_transgender_persons.pdf.

CHAPTER 5

1. Meghan Kallman and Josephine Ferorelli, "On the Future: A Harsh Climate for Motherhood," in *Motherhood in Precarious Times,* edited by Anita Dolman, Barbara Schwartz-Bechet, and Dannielle Joy Davis, 17–32 (Ontario, Canada: Demeter Press, 2018).

2. Carol Hanisch, "The Personal Is Political," 1969, https://webhome .cs.uvic.ca/~mserra/AttachedFiles/PersonalPolitical.pdf (accessed June 1, 2022).

3. These testimonies represent the voices of a sampling of house party participants who posted their recorded thoughts testimonies to Conceivable Future's website between 2017 and 2018. They stem from a fairly narrow range of experience, most of them coming from young white women in their twenties and thirties.

4. Note: These testimonies were recorded well before the Supreme Court's *Dobbs* decision repealed Constitutionally protected abortion access. Had *Dobbs* been decided prior to the testimonies, they may have been markedly different.

5. Kallman and Ferorelli, "On the Future," 22.

6. Rickie Solinger, *Pregnancy and Power: A Short History of Reproductive Politics in America* (New York: NYU Press, 2007).

7. Jade Sasser, *On Infertile Ground: Population Control and Women's Rights in the Era of Climate Change* (New York: NYU Press, 2018).

8. Thomas Robertson, *The Malthusian Moment: Global Population Growth and the Birth of American Environmentalism* (New Brunswick, NJ: Rutgers University Press, 2012).

9. Kallman and Ferorelli, "On the Future," 26.

CHAPTER 6

1. Intergovernmental Panel on Climate Change (IPCC), "Summary for Policymakers," in *Global Warming of 1.5°C. An IPCC Special Report on the Impacts of Global Warming of 1.5°C above Pre-Industrial Levels and Related Global Greenhouse Gas Emission Pathways, in the Context of*

Strengthening the Global Response to the Threat of Climate Change, Sustainable Development, and Efforts to Eradicate Poverty, 2018, edited by V. Masson-Delmotte, P. Zhai, H.-O. Pörtner, et al., www.ipcc.ch/sr15/chapter/spm/.

2. IPCC, "Summary for Policymakers" (emphasis added).

3. John Bowden, "Ocasio-Cortez: 'World Will End in 12 Years' if Climate Change Not Addressed," *The Hill,* January 22, 2019, https://thehill.com/policy/energy-environment/426353-ocasio-cortez-the-world-will-end-in-12-years-if-we-dont-address/.

4. Rebecca Zandbergen, "Q&A: Millions of Kids Will Walk Out of Class Today to Protest Climate Change," *CBC News,* May 3, 2019, www.cbc.ca/news/canada/london/london-ontario-climate-strike-emma-lim-1.5121373.

5. Jessica Murphy, "Hundreds of Thousands Join Canada Climate Strikes," *BBC News,* September 28, 2019, www.bbc.com/news/world-us-canada-49856860.

6. Lim quoted in John Bacon, "No Future, No Children: Teens Refusing to Have Kids until There's Action on Climate Change," *USA Today,* September 19, 2019.

7. Caitlin Stall-Paquet, "The Women Pledging Not to Have Kids until Meaningful Action on Climate Change Is Made," *Elle,* March 9, 2020, www.ellecanada.com/culture/society/the-women-pledging-not-to-have-kids-until-meaningful-action-on-climate-change-is-made.

8. Extinction Rebellion (XR), https://rebellion.global/ (accessed May 1, 2022).

9. Pepino as quoted in Simon Doherty, "The Activists Going on 'Birth Strike' to Protest Climate Change," *Vice,* March 13, 2019, www.vice.com/en/article/wjmkmz/the-activists-going-on-birth-strike-to-protest-climate-change.

10. Blythe Pepino, "#BirthStrike Mission Statement," #BirthStrike Tumblr, 2019, now defunct.

11. Jessica Gaitán Johannesson, "Why the 'Overpopulation Problem' Is an Issue of White Supremacy," gal-dem, January 16, 2023, https://gal-dem.com/overpopulation-problem-white-supremacy-climate-crisis/.

12. BirthStrike followers as quoted in Doherty, "Activists Going on 'Birth Strike' to Protest Climate Change."

13. Blythe Pepino, "How Can Global #BirthStrike Tackle, not Promote, Discrimination and Oppression While We Continue to be a Growing Public Movement?," BirthStrike statement, 2019.

14. As quoted in Heather McMullen and Katherine Dow, "Ringing the Existential Alarm: Exploring BirthStrike for Climate," *Medical Anthropology* (June 15, 2022): 1–15, DOI:10.1080/01459740.2022.2083510.

15. Jade Sasser, "From Darkness into Light: Race, Population, and Environmental Advocacy," *Antipode* 46, no. 5 (2014): 1240–1257.

16. Betsy Hartmann, "The Greening of Hate: An Environmentalist's Essay," in *Greenwash: Nativists, Environmentalism & the Hypocrisy of Hate* (Montgomery, AL: Southern Poverty Law Center, 2010).

17. Sarah Manavis, "Eco-fascism: The Ideology Marrying Environmentalism and White Supremacy Thriving Online," *New Statesman*, September 21, 2018, www.newstatesman.com/science-tech/2018/09/eco-fascism-ideology-marrying-environmentalism-and-white-supremacy.

18. Jordan Dyett and Cassidy Thomas, "Overpopulation Discourse: Patriarchy, Racism, and the Specter of Ecofascism," *Perspectives on Global Development and Technology* 18 (2019): 205–224.

19. Cassidy Thomas and Elhom Gosink, "At the Intersection of Eco-Crises, Eco-Anxiety, and Political Turbulence: A Primer on Twenty-First Century Ecofascism," *Perspectives on Global Development and Technology* 20 (2021): 30–54.

20. McMullen and Dow, "Ringing the Existential Alarm."

21. Pepino, *End of BirthStrike Statement.*

22. Pepino, *End of BirthStrike Statement.*

23. Jade Sasser, *On Infertile Ground: Population Control and Women's Rights in the Era of Climate Change* (New York: NYU Press, 2018).

24. Jessica Gaitán Johannesson, *The Nerves and Their Endings: Essays on Crisis and Response* (Melbourne & London: Scribe, 2022), 153.

CONCLUSION

1. Laura Briggs, *How All Politics Became Reproductive Politics: From Welfare Reform to Foreclosure to Trump* (Oakland: University of California Press, 2018).

2. Anna Louie Sussman, "The End of Babies," *New York Times*, November 16, 2019, www.nytimes.com/interactive/2019/11/16/opinion/sunday/capitalism-children.html.

3. Kath Weston, *Families We Choose: Lesbians, Gays, Kinship* (New York & Oxford: Columbia University Press, 2019).

4. Ruha Benjamin, "Black AfterLives Matter: Cultivating Kinfulness as Reproductive Justice," in *Making Kin Not Population* (Chicago: Prickly Paradigm Press, 2018).

5. Kim TallBear, "Making Love and Relations beyond Settler Sex and Family," in *Making Kin Not Population* (Chicago: Prickly Paradigm Press, 2018).

6. Jennifer N. Fink, *All Our Families: Disability Lineage and the Future of Kinship* (Boston, MA: Beacon Press, 2023).

7. Dean Spade, *Mutual Aid: Building Solidarity during This Crisis (and the Next)* (London: Verso Press, 2020).

8. Nina Jackson Levin, Shanna K. Kattari, Emily K. Piellusch, and Erica Watson, "'We Just Take Care of Each Other': Navigating 'Chosen Family' in the Context of Health, Illness, and the Mutual Provision of Care amongst Queer and Transgender Young Adults," *International Journal of Environmental Research and Public Health* 17, no. 19 (2020): 7346, https://doi.org/10.3390/ijerph17197346.

9. Dale Dominey-Howes, Andrew Gorman-Murray, and Scott McKinnon, "Queering Disasters: On the Need to Account for LGBTI Experiences in Natural Disaster Contexts," *Gender, Place and Culture: A Journal of Feminist Geography* (2013): DOI:10.1080/0966369X.2013.802673; Charlotte D'Ooge, "Queer Katrina: Gender and Sexual Orientation Matters in the Aftermath of the Disaster," in *Katrina and the Women of New Orleans*, edited by Beth Willinger, 22–24 (New Orleans: Tulane University Press, 2008); and Leo Goldsmith, Vanessa Raditz, and Michael Méndez, "Queer and Present Danger: Understanding the Disparate Impacts of Disasters on LGBTQ+ Communities," *Disasters* 46, no. 4 (2022): 946–973.

10. Harriet E. Ingle and Michal Mikulewicz, "Mental Health and Climate Change: Tackling Invisible Injustice," *Lancet Planet Health* 4, no. 4 (2020): e128–e130.

11. Danielle Carr, "Mental Health Is Political," *New York Times*, September 20, 2022, www.nytimes.com/2022/09/20/opinion/us-mental-health-politics.html.

Index

170   INDEX

reproductive resistance, 20, 26–28,
36–42, 44, 98, 108; of BirthStrike,
117–118, 133; among Black enslaved
women, 26; climate change and, xii–
xvii, 61–62; defined, 23; of Millennials
and Generation Z, 119; of #NoFu-
tureNoChildren, 117–118, 132–133;
reproductive autonomy and, 57, 133;
social context of, 33–34; to state policy, 57–58
Roe v. Wade, xvi; the kid question and, 7;
repeal of, 7, 51–52

Sanger, Margaret, 27
See Change Institute, 43–44
Sherrell, Daniel, 11–12
SisterSong Women of Color Reproductive
Justice Collective, 108
social media: parenthood and, xv; use by
BirthStrike, 123, 125, 127–128
solastalgia, 12
Solnit, Rebecca, 18
sterilization, forced, 105, 130–131, 136
sustainability, environmental, xvi

"talk, the" 42
Thunberg, Greta, 119, 137
trauma: climate change and, 77; of racial
inequality, 82
Trump, Donald, 40, 115

violence: against Black people, xvi, 66,
82–83; colonial, 126; against

Indigenous women, 66–67; racist,
9, 50
vulnerability: of Black children, 81–82,
83; of Black people, 78, 81–83; class,
77–78; climate emotions and, 93;
Covid-19 pandemic and, 54; of Generation Z, 83; the kid question and,
68; of people of color, 66–68; poverty
and, xv; of pregnant women, 54;
racial, 71–72, 81–82; racism and, xv,
77–78

white privilege: disaster recovery and,
77–78; eco-anxiety and, 4–6
white supremacy: ecofascism and, 10,
128; population control and, 130
white women: the kid question and, 100–
103; motherhood, 105
wildfires, 111–112; Bobcat Fire (Pasadena,
CA), xi–xiii; Camp Fire (Paradise, CA),
40–41; fire insurance, 96–97; health
effects, 76; Woolsey Fire, 22, 88
"women": use of term, 147n1
women of color: abortion access of,
52–53; barriers to reproductive health
care, 56; preterm births among,
53–54; reproductive justice and,
10–11, 47, 55, 106–109, 136. *See also*
people of color
Wray, Britt, 5–6, 35, 69

Zero Population Growth (environmental
group), 27

Founded in 1893,
UNIVERSITY OF CALIFORNIA PRESS
publishes bold, progressive books and journals
on topics in the arts, humanities, social sciences,
and natural sciences—with a focus on social
justice issues—that inspire thought and action
among readers worldwide.

The UC PRESS FOUNDATION
raises funds to uphold the press's vital role
as an independent, nonprofit publisher, and
receives philanthropic support from a wide
range of individuals and institutions—and from
committed readers like you. To learn more, visit
ucpress.edu/supportus.